Sir Douglas Galton

Observations on the Construction of Healthy Dwellings

Namely, Houses, Hospitals, Barracks, Asylums, etc.

Sir Douglas Galton

Observations on the Construction of Healthy Dwellings
Namely, Houses, Hospitals, Barracks, Asylums, etc.

ISBN/EAN: 9783337163075

Printed in Europe, USA, Canada, Australia, Japan

Cover: Foto ©berggeist007 / pixelio.de

More available books at **www.hansebooks.com**

OBSERVATIONS

ON

THE CONSTRUCTION

OF

HEALTHY DWELLINGS

NAMELY

HOUSES, HOSPITALS, BARRACKS, ASYLUMS, ETC.

BY

SIR DOUGLAS GALTON, K.C.B.

Late Royal Engineers, Hon. D.C.L., LL.D., F.R.S., Hon. Member Inst. C.E., M.I.M.E.
F.S.A., F.G.S., F.C.S., F.L.S., F.S.I., F.R.G.S.
Formerly, Secretary Railway Department Board of Trade
Assistant Inspector-General of Fortifications
Assistant Under Secretary of State for War
Director of Public Works and Buildings
&c. &c.

SECOND EDITION REVISED

Oxford

AT THE CLARENDON PRESS

M DCCC XCVI

PREFACE TO THE FIRST EDITION

—·—

THE researches of the physiologist and of the medical man into the laws which govern the prevalence of diseases have enabled them, by the gradual accumulation of information, to lay down the principles upon which the healthy construction of houses should rest. It is the duty of the architect, the builder, the engineer, and the surveyor to apply these principles, and their correct application is as essential to the efficient construction of a dwelling as is the quality or strength of the materials which are used to build the dwelling.

The object which the author has in view in this treatise is to present in a condensed form a short *résumé* of the very scattered information which exists bearing on the construction of healthy dwellings, whether houses, hospitals, barracks, asylums, or prisons: this information was embodied originally in a series of lectures delivered to the officers of Royal Engineers in their educational establishment at Brompton Barracks, Chatham. The author has expanded the information thus collected into this volume, which contains partly an enunciation of principles, and partly brief sketches to elucidate the application of these principles. The field embraced is so vast that it was beyond the scope of this work to enter into the details which a complete text-book would entail; and those whose business calls upon them to apply practically the principles of sanitary construction, must seek the further information they may need from one or other of the numerous separate treatises already in existence on each subject.

There is necessarily little that is new in this work. It contains much that the author has derived from personal experience, as well as from the opportunities he has had of becoming acquainted with the progress of sanitary construction in many European countries, in the United States of America, and in our Indian dominions; much also has been collected from published books; and he trusts that the several authors of these books will accept the acknowledgement of their works in this preface, and pardon him for not having invariably furnished references to the extracts in the body of the work. The following, among these works, may be usefully referred to by those who would pursue the subject further; as well as by architects, builders, engineers, and surveyors, who are concerned in the application of details, viz:—The Reports of the Barrack and Hospital Improvement Commission, the Reports of the Commission on Cubic Space in Workhouses, and Pollution of Rivers, Army Sanitary Commission, Indian Sanitary Reports, Reports of Board of Health, Reports of Privy Council, Reports of Local

Government Board, Dr. Parkes on Hygiene; the
works of General Morin, M. Tresca, M. Joly, Dr. De
Chaumont, F. Sander (Leipzig), R. Rawlinson, Richard-
son, Baldwin Latham, Humber on Water Supply,
Bailey Denton on Sanitary Engineering, Dr. Edward
Smith's Manuals, Hellyer on Healthy Houses, Teale's
Sanitary Defects, Buchan's Plumbing, Hood on Warm-
ing and Ventilation, Leeds on Ventilation (New York),
Edwards on Fireplaces, Rankine's Tables, Balfour
Stewart on Heat, Box on Heat, Péclet sur la chaleur,
Symons' Rainfall, Buchan's Meteorology, Ansted's
Geology, Tidy's Chemistry, Hurst's Architectural Sur-
veyor's Handbook, Dr. Angus Smith's works; also
numerous papers on Sanitary Science in the Proceedings
of the Royal Society, Chemical Society, Institutions of
Civil and Mechanical Engineers and Surveyors, Society
of Arts, etc., and many others.

PREFATORY REMARKS TO SECOND EDITION

—◦—

SIXTEEN years have now elapsed since the First Edition of my work on Healthy Dwellings was published.

I have only to add to my original Preface to that Edition, that I have been very glad to embrace the opportunity which the demand for a Second Edition affords me, of embodying into the text many of the developments produced by the great progress in the Science of Hygiene.

DOUGLAS GALTON.

CONTENTS

—•—

CHAPTER I.

LIST OF DIAGRAMS

———◦———

TABLES

————

THE CONSTRUCTION

HEALTHY DWELLINGS.

—••—

CHAPTER I.

PRELIMINARY OBSERVATIONS.

HYGIENE is generally defined as the science of the preservation of health—that is to say, the science of counteracting the influences injurious to health which arise in the surroundings in which man is placed. These influences may be classed as—

1st. Physical; such as conditions of heat, light, electricity, sound, &c.

2nd. Chemical; such as conditions of soil, air, water, food.

3rd. Biological or individual; such as conditions of race, sex, age, inheritance, constitution, temperament.

4th. Social; such as conditions of profession, family, or nation.

Hygiene, in its wider sense, includes a study of all these.

The present treatise is limited to a consideration of those conditions included in the first two classes, in which the architect and the engineer are more especially concerned.

The statistician and the physiologist supply the data which enable us to determine what is a healthy condition of life; and their researches show that the majority of diseases, and the greater part of the low health, which prevail in any country, arise from causes which are within man's control.

For instance, epidemic diseases are observed to occur in very different degrees of intensity at different periods, amongst groups of population exposed to certain unhealthy conditions. Sometimes they take the form of pestilences, and immediately afterwards, the conditions remaining the same, they subside and all but disappear, again to renew their ravages at some future periods.

The biologist assures us that certain diseases, such as cholera and diphtheria, for instance, are connected with micro-organisms which have obtained access to our bodies through the air we breathe, or through our food or water; but he has not yet explained to us the conditions under which they affect different races in different degrees, nor why these epidemics break out at one time and not at another. There are however certain well-defined conditions which influence materially, not only their actual intensity, but also their frequency.

Thus intermittent fever was observed to disappear from places which formerly suffered from it, after drainage of the soil and improved cultivation. By cleanliness, fresh air, and diminished crowding, the very worst forms of pestilential fever, which used to commit ravages similar to those of the plague, disappeared entirely from English gaols.

The breathing of foul air contaminated by the breath of other persons appears to be the special agent which develops consumption, and diseases of that class.

Consumption and tubercular disease used to be rife in the British army, because barrack rooms were crowded and unventilated, and the atmosphere close or foul during the hours of sleep, when the system is more peculiarly predisposed to its effects. Out of such an atmosphere, which the men had been breathing night after night, they were taken and exposed on guard to wet and cold, and the disease soon developed itself. The cessation of these preventible causes has led to a great diminution of the disease; but it is still far too prevalent in the army.

Zymotic diseases, namely fevers, diarrhoea, cholera, dysentery, &c., are most intensely active where there is over-crowding, and the repeated breathing of air already breathed, such air being further contaminated by moisture and exhalations from the skin ; equally poisonous are emanations proceeding from animal excretions, or from decaying vegetable matter, together with moisture, the want of drainage, and a foul state of latrines, urinals, cess-pits, and manure heaps. Moreover, cholera and dysentery are intimately connected with the condition of the water supply : while an epidemic prevails, the question whether a given population shall suffer or escape may almost be predicted from a careful analysis of the drinking water.

Hence, the causes of the deteriorated health which lie more especially within the scope of the present work, arise from poisons in the soil we live on, the air we breathe, or the water we drink—emanating from decomposition, which is the result of the previous occupation of the locality by some form of animal or vegetable life.

When, therefore, the degree of health in a community, a household, or an individual, falls below the ascertained standard of good health, it is the duty of every individual in that community to seek out the removable causes in operation which are injurious to health, and to remove them.

Practical sanitary science is thus embodied in the words, pure air, pure water—these conditions include a pure subsoil. Where these exist, the highest degree of health of which the human constitution is capable may be anticipated. These conditions imply a healthy site, a healthy dwelling, and the rapid removal from its vicinity of everything that is liable to putrefy.

Whilst it is for the physiologist and chemist to point out the prevalence of the poison, it is the function of the engineer and architect to devise means for obviating its baneful effects.

CHAPTER II.

THE healthiness of a site depends, not only on the position itself, but also on what lies around it.

The selection both of the position for, and the form of, a dwelling must largely depend on the climate; that is to say, on temperature, on rainfall, on moisture of soil, on the nature and prevalence of winds.

The engineer may modify the conditions and temperature of the soil, and diminish atmospheric damp by drainage; he may alter the moisture and temperature of the air by planting or removing forests; he may produce changes in the immediate surroundings of a locality; but those general climatic conditions of a country which are due to position on the globe and to the vicinity of seas or continents are beyond the control of an engineer.

1. *Temperature of Atmosphere.*

In classifying the temperature of a country, it may be said that when the mean temperature of the locality lies between the thermal equator and the isothermal line of 80° Fahrenheit, the climate is torrid; between the isotherm of 80° and that of 60° the climate is hot; between the isotherm of 60° and that of 40° the climate is temperate; between the isotherm of 40° and 24° Fahrenheit, the climate is cold. A greater degree of cold represents a polar climate.

But temperature decreases with the altitude of a district

above the sea. This decrease is, however, subject to many variations dependent upon latitude, situation, dampness and dryness of atmosphere, upon the surroundings of the locality, whether water, or forest, or desert, or mountains, and even upon the season of the year and hour of the day. A general rule, subject to these variations, has been laid down as follows ; viz. each 300 feet of height above the sea lowers the temperature about $1°$ Fahrenheit.

Similarly, in passing from the equator towards the pole, the mean temperature may be said to be lowered by about $9°$ Fahrenheit for every $10°$ of latitude. A greater variation of climate, however, than that due to latitude arises from other causes ; for instance, the relative proximity of a place to the ocean, the temperature of which prevents an extreme degree of either cold or heat in countries bordering thereon ; also, the effect of mountain ranges, which deprive the winds which pass over them of their moisture, and allow of a more complete radiation of heat from the ground, during the long winter nights, and thus produce an intense degree of cold.

The position of a site as compared with the level of the adjacent country affects its temperature. For instance, when the air in contact with declivities of hills and rising ground becomes cooled by the ground, the cold air will flow down the sides of hills into the valleys, displacing the warmer air and forming as it were pools of cold air. Thus rising ground is never exposed to the full intensity of the cold.

It is beyond the scope of this treatise to enter fully into the subject of climatic conditions and temperature. It may however be observed, that the death-rate in any locality has been stated to increase with the increase in the difference of the mean temperature of January and July of that locality.

One of the principal dangers to health in torrid and hot climates lies in the fact that heat and moisture especially favour the decomposition of animal and vegetable matter; and these climates are, from this cause, liable to be eminently unhealthy. A high range of mean temperature is therefore an important element in tropical and subtropical climates. In the temperate and cold latitudes the conditions of decomposition are not so intense ; and where they exist they are more easily controlled.

2. *Temperature of Soil.*

Daily changes of temperature do not affect the soil to a greater depth than three feet, varying with the daily range of temperature.

The annual variation is dependent on the conductivity and specific heat of the soil, but it does not penetrate below 40 feet, and below 24 feet it is very small. The mean temperature of the soil follows slowly the mean temperature of the air. The highest annual temperature of the trap rocks at Calton Hill, Edinburgh, at a depth of 24 feet, takes place about the 4th of January, and the greatest cold about the 13th of July. At Greenwich, which is on the tertiary gravels, the highest temperature at a depth of 25.6 feet occurs on the 30th of November, and the lowest on the first of June. At a depth of 12.8 feet the highest temperature occurs on the 25th of September, and the lowest on . the 27th of March. For all practical purposes the temperature in the soil at a depth of from six to eight feet, may be said to be practically fixed all the year round, because it follows so slowly the summer and winter changes that it never attains summer heat or winter cold.

The temperature of the earth increases with the depth. This rate of increase of temperature varies in different geological formations.

In the Paris basin it has been estimated at 1° Fahrenheit for every 55 feet of depth. In England it has been stated at 1° for every 54½ feet. It is also stated that it varies with the latitude, being lower in the higher latitudes, and higher towards the equator.

The mean rate of increase over the globe may be approximately assumed at $1°$ Fahrenheit for every 50 feet in depth.

The internal heat exercises, under certain conditions, an influence over the mean temperature of the surface soil of a locality.

Where the rainfall is tolerably evenly divided over the year, the average annual temperature of the soil will be that of the climate of the locality. But in countries with distinct wet and dry seasons, the mean temperature of the soil will not necessarily be the same as that of the mean temperature of the air. Snow, being a bad conductor, prevents the passage of the heat from the earth into the air, and thus in countries where snow lies for some time on the ground, the mean temperature of the earth exceeds that of the air.

The temperature of water in permanent springs is necessarily derived from the subsoil line of fixed temperature, and, except in the cases just alluded to, it will not be found to vary more than $1°$ or $2°$ from the mean temperature of the locality; therefore the temperature of a permanent spring may be assumed to afford a certain guide to the mean temperature of a district.

Thus, in England, the permanent springs range in temperature from $49°$ to $51°$, the mean annual temperature being $50°$. In India the springs will be found to vary in different parts, according to the temperature of the locality ; in some instances they attain a temperature of from $70°$ to $80°$.

But whilst the mean temperature of the ground depends on the climate, soils have a very varying conducting capacity for heat ; loam, clay and rocks are better conductors than sand, and by allowing the sun's heat to pass more rapidly downwards, do not become heated to so high a degree.

The conducting capacity of the soil has a very important bearing upon the comfort, if not upon the health, of those who live upon it.

The following table shows the relative power of soils to

retain heat : sand being the worst conductor, 100 is allotted
to it as the standard :—

Sand, with some lime	100.0	Clayey earth	68.4
Pure sand	95.6	Pure clay	66.7
Light clay	76.9	Fine chalk	61.8
Gypsum	73.2	Humus	49.0
Heavy clay	71.11		

The great retentive power of the sands is thus evident, and
the comparative coldness of the clays and humus.

Under exposure to the sun's rays, herbage lessens the
absorbing power of the soil, and radiation is more rapid
from it, because a portion of the heat is lost by the evaporation which goes on from the pores of plants, and the leaves
are rapidly robbed of their heat by the adjacent air.

Changes of temperature take place slowly in trees as
compared with the temperature of the air. Trees acquire
the maximum temperature after sunset, whilst the maximum
temperature of the air occurs between 2 and 3 p.m. Hence
the influence of trees is to make the night warmer and
the days colder ; and the heat is more evenly distributed
over the twenty-four hours in countries covered with vegetation than in those free from it.

Evaporation goes on slowly under trees ; but the vapour,
not being so liable to be removed by the wind, accumulates
among the trees. Hence, whilst forests diminish evaporation,
they increase humidity, and they keep the summer temperature lower, and the winter temperature higher than it would
be without them. For these reasons, forests may act in the
same way as a range of hills, to increase rainfall in the
summer by causing condensation in the case of warm moist
winds.

3. *Moisture of Soil.*

The condition which, more than any other, governs the
healthiness of the soil, is the relation which the ground air,

or air in the soil, bears to the ground water; that is to say, the presence or absence of moisture in the soil.

As a rule, rainfall is the parent of the ground water. Hence, the moisture in the soil, or ground water, depends upon the amount and mode of incidence of the rainfall. and of the facilities for its removal.

Rain varies greatly, both as regards frequency and rate of fall. In some places it never rains. The heaviest recorded average annual rainfall is said to be 600 inches. This occurs on the Khasia Hills, which rise abruptly opposite the Bay of Bengal, and are separated from it by 200 miles of swamps. As much as 700 inches have fallen in a year. Five-sixths of the quantity of rain falls in about half the year; 264 inches have been recorded to have fallen in the month of August alone, and 30 inches to have fallen in one period of 24 hours.

The basin of the Indus contains districts where the rainfall is sometimes as low as six inches in the year.

In England the average rainfall is 32 inches, but the average gives a wrong impression of the condition of different parts of the country.

In the west of Great Britain and Ireland, in the immediate neighbourhood of hills, the average rainfall is above 75 inches; and in some localities 150 inches have been observed: in some years it is even higher. In the east of Great Britain, from 20 to 28 inches of rain falls. The amount of rainfall is affected by proximity to the sea, as well as by mountain ranges and hills; from these latter, and other causes, it varies materially at places a short distance apart, and therefore each locality must be considered separately.

The average of a series of years is not what an engineer must look to. He has to deal with the maximum or minimum. In questions of water supply, the minimum of the yearly fall is what must influence his calculations. But in questions of the removal of water, he must look to the maximum, not of the yearly fall only, but to the greatest amount which may fall in a limited period. The driest years test water supply ; the wettest years test works.

In the 24 hours ended 9 a.m., on the 15th of July, 1875, there was recorded as having fallen at Newport, 5.33 inches ; Tintern, 5.31 inches ; Cardiff, 4.7 inches.

There are other instances on record of rain having fallen in England at the rate of four inches in an hour on a limited area. In fact the rain in this country occasionally falls as heavily as in India, but a heavy fall does not last for so long a time.

It has been estimated from observations in this country that the maximum annual rainfall exceeds by one-third, and the minimum annual rainfall is less by one-third, than the mean rainfall of a series of years. It has also been observed that the average annual rainfall of three consecutive dry years amounts to 80 per cent. of the mean annual rainfall of a series of years.

The time of the year at which the rain falls, and the resulting effects on the air and the soil, is a point of great sanitary importance.

In dry seasons the re-evaporation is rapid. In India, rain may fall at such intervals in a dry season as to allow of entire re-evaporation. In wet seasons water falls on water and flows off in floods.

Experiments would appear to show that in this country, on an average, nearly two-thirds of the mean annual rainfall goes back at once into the atmosphere as evaporation ; but these experiments were chiefly made on parts of the country where the rainfall does not materially vary from the mean annual rainfall. A further portion evaporates slowly from the soil, or is absorbed by vegetation. The water which is not evaporated, or absorbed by vegetation, percolates into the soil, to flow out in springs, streams, and rivers, to the sea,[1]

[1] The Upper Thames basin contains 3676 square miles ; the mean daily flow of water over Teddington Lock during ten years, added to the water pumpe

whence it again passes into the atmosphere, and returns as rainfall or as dew. For it is evident that if the humidity of the atmosphere be assumed to be constant for an average of years, the evaporation over the whole globe must equal the rainfall.

The proportion of evaporation and percolation in different soils varies—

1. With the time of year in which the rain falls.
2. With the quality of the soil, with its capacity for heat, and with the character and extent of the vegetation with which it is covered.

In England, on an average of years, the spring is the driest and the autumn the wettest part of the year. The driest months are March and April, the two wettest months are October and November.

The greatest percolation takes place after a wet period, when the soil becomes saturated. In the summer there is scarcely any percolation. A series of experiments on percolation in England, extending over 14 years, showed that in five of the years there was no percolation during a continuous period of seven months; and that in one year only, viz. 1860, did percolation take place every month.

In a series of experiments by Mr. Dickenson, it appeared that from April to September, 93 per cent. of rainfall was evaporated, and 7 per cent. absorbed; whilst from October to March, 25 per cent. was evaporated and 75 per cent. absorbed.

Great percolation follows the thawing of snow, and the greatest percolation is due to frequent small falls of snow.

The nature of the soil affects the percolation. Through sand the percolation is great. The evaporation from sand in this country has been shown by experiment to be 16

into London by the Water Companies, was nearly 140,000,000 cubic feet; the average rainfall in the Upper Thames basin during those years was 29·5 inches = 690,000,000 cubic feet per diem. The proportion which the water brought down by the river bore to the rainfall during the period was as 1 : 4.9; but a portion flowed as an underground river in the porous strata in which the river is situated.

per cent., and the percolation 84 per cent., whereas with clay and loam, the percolation was found to be 27, and the evaporation 73 per cent.[1]

In hot climates the relative power of various soils to retain heat would alter these proportions ; for instance, the evaporation in such cases would be large from sand if unprotected by herbage.

The presence or absence of vegetation exercises an important influence on percolation. When vegetation is rapid, as in the case of growing crops in the spring, it arrests percolation ; when the ground is covered with forests, the moisture is retained nearer the surface.

[1] The following Table gives the result of various experiments on percolation through different soils :—

Authority.	Material.	Duration of Experiment.	Percolation per cent.	Evaporation per cent.
Dalton	Earth / Surface grass	3 years	25·0	75·0
Dickenson . . .	Gravelly loam / Surface grass	8 ,,	42·5	57·5
Maurice	Earth	2 ,,	39·0	61·0
Gasparin . .	Earth	2 ,,	20·0	80·0
Rister	Impervious subsoil / Sand, cropped	2 ,,	30·0	70·0
	Mean . .	——	31·3	68·7
Greaves	Loam, gravel, and sand, turfed	22 ,,	26·6	73·4
	Sand	14 ,,	83·2	16·8
Laws & Gilbert . .	Loam, clay subsoil, built in 20″ deep	5 ,,	36·8	63·2
	Loam, clay subsoil, built in 40″ deep	5 ,,	36·0	64·0
	Loam, clay subsoil, built in 60″ deep	5 ,,	28·6	71·4
Dickenson & Evans	Surface soil at Nash Mills .	25 ,,	31·0	69·0
Evans	Soil	20 ,,	22·0	78·0
	Chalk	20 ,,	36·0	64·0

Dr. Ebermeyer found at Salzburg that the percolation was

in May, 25·2 ⎫
June. 53·1 ⎪
July, 23·4 ⎬ per cent. less through turf than through bare earth,
Aug., 29·2 ⎪
Sept., 72·7 ⎭

and that the difference was least in January.

Ebermeyer's experiments. moreover, showed that in the summer half year, forest soil is moistest, bare open ground less moist, turf driest.

Forests retain the moisture, protect the soil, and in mountainous countries retard the flow of torrents. Turf produces to some extent the same effect.

The power of retention of water in the soil exercised by the planting of trees was exemplified in the Island of Ascension. That island formed a convenient point for ships to call at for obtaining water on their way home from the East Indies. It was a barren rock, to which formerly the water had to be conveyed in ships. Some years ago, trees were planted on the Island These have thriven, and now the rain which falls, instead of passing away at once into the atmosphere by evaporation, is retained in a sufficient quantity to fill tanks with water for the supply of the ships which call there. The problem of the water supply of some others of our insular possessions might be solved by similar proceedings.

On the other hand, all growing vegetation evaporates a large quantity of water. In order to form 1 lb. of woody fibre, a plant evaporates 200 lbs. of water; consequently a country covered with forest evaporates an enormous quantity of water out of the soil, in order to produce a growth of timber.

When the rainfall has penetrated from 12 inches to 2 feet into the ground, the loss from evaporation is comparatively small; but it varies with the nature of the soil. In the chalk formation, water will rise by capillary attraction from the level at which the chalk is saturated, to a considerable height above that point; while a bed of sand will be dry at the height of about a foot above the water standing in it.

So long as there is water sufficiently near the surface of the soil to keep it moist by attraction, evaporation will continue. Clay and similarly retentive soils do not give off vapour as copiously as free open soils; therefore a given quantity of moisture will occupy a longer period in passing off these soils than is the case with free soils under similar conditions.

The capacity of soils to retain water varies greatly. Impermeable granite or marble will hold about a pint of water per cubic yard. Pure sand will hold 40 or 50 gallons, or about 17 per cent. of its weight when dry, and the ordinary red sandstone rock 27 gallons per cubic yard, or from 7 to 8 per cent. of its weight when dry. London clay will hold 50 per cent. of its own weight when dry, Oxford clay about 30 per cent. Experiments made by Mr. B. Latham in 1879, for the Royal Agricultural Society, gave the following percentages of water by weight which surface soils selected from different localities will absorb, viz. open gravel from 9 to 13 per cent.; gravelly surface-soil 48 per cent.; light sandy soils from 23 to 36 per cent.; loamy soil 43 per cent.; yellow marl subsoil 25·9 per cent.; stiff land and clay soils from 43·3 to 57·6 per cent.; sandy and peaty soils from 61·5 to 80 per cent.; peat 103 per cent. The conditions of two localities may thus vary greatly, although there may be an apparent general similarity in the soils.

Under-draining facilitates the passage of the water from the surface into the ground; and a smaller evaporation and greater percolation takes place in drained lands.

A drained field will consequently have a temperature as much as 6° or 7° Fahrenheit higher than an adjacent undrained field. The cause is obvious. To convert water into vapour absorbs 960° Fahrenheit of heat from its vicinity;

thus each cubic foot of water evaporated will lower the temperature of something like 3,000,000 cubic feet of air 1°. The lower the water in the soil, the less the evaporation, and the warmer the adjacent air.

The discharge of underdrains in a free soil of chalk, gravel, and sand was found, by Mr. Bailey Denton, to be 2¾ times as rapid as that from underdrains in a clay soil; the drains in the clay being 25 feet apart ; and those in the free soil placed at irregular intervals widely apart.

The rate at which a soil allows of the percolation of water regulates the distance apart at which underdrains should be placed, for the purpose of lowering the subsoil water in land. In free soils a single drain will lower the water for a large area ; in clay soils numerous drains are necessary. The discharge from open soils is more regular than from clay ; although clay soils often give out a large proportion of the rainfall immediately after it occurs. Barometric pressure may affect the discharge from drainage outlets ; an increased discharge has been observed to follow a fall in the barometer without any fall of rain on the surface.

In a country where the proportions of mean annual rainfall vary so much as they do in Great Britain at comparatively short distances apart, the areas of heavy rainfall have an important bearing on water supply. On these areas the rainfall is more continuous, and the actual amount evaporated will therefore be less than in the drier parts of the country. Hence the proportion to the total rainfall, of rainfall which can be collected from these areas, will be many times greater than what could be obtained under the most favourable circumstances in the drier parts of the country.

4. *Aeration of Soil.*

The level of water in the subsoil regulates the amount of ground air.

Air permeates the ground, and occupies every space not

filled by solid matter, or by water. Thus, it is the same thing to build on a dry gravelly soil, where the interstices between the stones are naturally somewhat large, as to build over a stratum of air. The air moves in and out of the soil in proportion to barometric pressure, and with reference to the wind. If there is much water in the soil, the air carries with it watery vapours, and is cold, and such a site is called damp.

The fact of this continual free passage of air in and out of the ground makes it important that, not only should the ground lived on be free from water, but that it should also be free from impurities. It would be just as healthy (indeed probably far healthier) to live over a pigstye than over a site in which refuse has been buried, or in which sewer water has penetrated, or over a soil filled with decaying organic matter ; thus, before building on any ground, its nature should be carefully examined.

It must, however, be remembered that in proportion as there is a free movement of air in the soil, so is the process of decay, and consequently the removal of decaying matter, more rapid. Louis Créteur, in his work 'Hygiene in the Battle Field,' gives his experience in disinfecting the pits where dead were buried near Sedan. The bodies were buried in chalk, quarry rubble, sand, argillite, slate, marl, or clay soils, and the work of disinfection lasted from the beginning of March till the end of June. In rubble the decay had taken place fully, but in clay the bodies were surprisingly well kept, and even after a very long time the features could be identified.

Experiments have demonstrated that there is a considerable quantity of carbonic acid in the ground : a result frequently of the decay of organic matter. Water which comes up from deep springs always contains carbonic acid, but it has been shown that the ground air frequently contains 50 per cent. more carbonic acid than the ground water ; it seems, therefore, that the latter is supplied with its carbonic acid from the ground air. Recent experiments in India have demonstrated that the carbonic acid in air drawn in a particular locality from a depth of three feet was one-half that obtained from a depth of six feet.

These experiments have also shown that there are considerable variations in the amount of carbonic acid present in the soil of localities in close proximity ; the amount was found to be doubled in a distance of 50 yards, with apparently similar soil. The processes going on in the soil at these two spots must have differed materially ; and if such processes affect health, persons inhabiting a building over one of these sites would be exposed to different hygienic conditions from persons living over the other.

These facts show the immense importance which the soil on which dwellings are placed exercises upon health, especially in cold or damp weather, when the air in the dwelling is warmer than the air outside ; for then the upward movement of this warmer air will draw in air to supply its place, from the ground on which the dwelling stands. This movement would be prevented if the whole surface under the dwelling were covered with an impervious material. But it is difficult to find a building material impervious to air.

The following table shows the volume in cubic feet per hour of air which passed through a square yard of wall surface of equal thicknesses built of the following materials, the pressure being obtained by a difference of temperature of $72°$ Fahrenheit inside, and $40°$ Fahrenheit outside :

Wall built of Sandstone . . . . 4·7 cubic feet.
    do.   Quarried Limestone . . . . . 6·5 ,, ,,
    do.   Brick . . . . 7·9 ,, ,,
    do.   Limestone . . . . 10 1 ,, ,,
    do.   Mud . . . . 14·4 ,, ,,

Concrete is the most convenient covering to be placed under a building ; it is not impermeable ; the lime in it will absorb the carbonic acid from the ground air for a certain time, until it is all converted into carbonate of lime, and

C

then its beneficial action would be diminished. Asphalte is more impermeable than concrete.

The amount of the air drawn from the soil may be checked by raising the floors above the surface of the ground, and affording free circulation of air from outside between the raised floor and the ground. In Burmah, the dwellings are raised on poles. In Italy and France it is usual to build a basement on open arches, which are used for wood or other stores, over which the dwelling is constructed.

So little, however, has the influence of ground air been appreciated in this country that it is only comparatively recently that the occupation of cellars with walls abutting on the soil, as habitations, has been prohibited ; and even within 20 years rooms in the basements of barracks have been used as barrack-rooms for soldiers. The plan of allowing the earth to rest against the walls of rooms in basements is still unfortunately common.

These considerations show the importance of forming a deep open area round a house, and carrying it below the level of the basement floor ; between the floor of any living room and the ground there should always be ventilation to the outer air. For similar reasons, a trench should always be dug round a tent. Apart from the advantage which this affords for draining off the water, the sides of the trench enable the atmospheric air to permeate freely to the ground immediately under the tent, especially when the tent stands on a gravelly soil.

Whilst a permanently low water level (say 15 feet) in the soil may be termed healthy, and a permanently high water level (say under 5 feet) may be termed unhealthy, a fluctuating water level is very unhealthy, especially when the fluctuations are rapid. The unhealthiness mainly shows itself when the level of the ground water falls.

Fever chiefly occurs in flooded districts when the floods have receded, and in some of the Indian districts where the earth surface is covered by water and secluded from the air there is no cholera ; but when the water falls and the surface of the country is drying up cholera reappears. This unhealthiness may be due to the decay of organic matter left by the receding water. Floods under these conditions afford for a time a cloak to the impurities upon and in the soil ; but when the water falls, it leaves the impurities behind it, which pollute both the surface and the water sources.

It is for these several reasons that it is desirable to keep the permanent level of water in the soil, where habitations are placed, as low as possible ; that is to say, to drain it, so as to allow the air to have free play in the soil. But where the ground water cannot be maintained permanently at a low level, then keep it at an even level.

Thus the presence or absence of moisture determines very much the degree of healthiness of soils. In any country, that area over which fogs appear soonest after nightfall should be avoided for camping, and should be drained before building. The water level of every camping ground should be examined by digging holes ; but a correct idea can be obtained as to where water is nearest the surface, from observing where the vegetation is greenest, where midges prevail in the day time, and where fogs appear soonest at nightfall.

Apart from its effect on the moisture of the soil, vegetation has an influence of its own on the healthiness of a site. Plants in respiration absorb oxygen and throw off carbonic acid, but the action of the cells in which the green matter termed chlorophyll is formed, which gives colour to vegetation, is to absorb carbonic acid and to eliminate oxygen.

This action is due to the sun's rays. Vegetation is thus beneficial, but when in the vicinity of dwellings it should always be vigorous, healthy, and green ; fallen leaves and decayed vegetation should be rapidly removed like other refuse from the vicinity of dwellings.

CHAPTER III.

DR. PARKES gives the following table to show the relative healthiness of various geological formations, based upon their relative permeability:

		Permeability of water.	Emanations into air.
I.	Primitive rocks, clay slate, millstone grit.	Slight.	None.
II.	Gravel and loose sands, with permeable subsoils.	Great.	Slight.
III.	Sandstones.	Variable.	Slight.
IV.	Limestones.	Moderate.	
V.	Sands, with impermeable subsoils.	Arrested by subsoils.	Considerable.
VI.	Clays, marls, alluvial soils.	Slight.	Considerable.
VII.	Marshes, when not peaty.	Slight.	Considerable.

Dr. Parkes states that cholera is unfrequent on granite, metamorphic, and trap rocks. Granite districts are usually sparsely peopled and the element of overcrowding as a cause of disease is absent. But a careful collation of facts shows that so far as cholera is concerned, the site and the geological formation are not the principal contributing causes: it occurs where there is a population in a filthy condition. And as a rule, local conditions modify the effect of soil and geological formation. For instance, granitic and other impermeable formations are termed healthy because the impurities, instead of passing into the soil, are carried off rapidly by rainfall; but if filth is allowed to accumulate they will be unhealthy.

Thus during the first visitation of cholera, one of the places which suffered most severely, owing to its filthy local condition, was Megavissey, on the granitic formation in Cornwall.

There was much sickness and mortality in 1859–60 at Hong Kong. The peninsula of Kowloon was selected as a sanatorium. It was of granite formation, freely exposed to the winds; it was reputed to possess every quality for health. Huts were built, and the troops were moved into them. They suffered severely from fever.

This arose from the disturbance in a tropical climate of the surface soil impregnated with decaying organic matter. Until soil of that nature has been opened and oxygenised, it is in the highest degree deleterious.

Brushwood is a source of danger near camps in a hot climate, but the immediate result of the removal of brushwood has been found to cause fever, owing to the disturbance of decaying organic matter occasioned thereby.

In cold countries, the clayey soils are cold; and as they prevent the percolation of rainfall, they are also damp, and favour the production of rheumatism and catarrhs; for these reasons sands are generally the healthier soils in this climate. Sand or gravel soils are, however, only healthy if kept entirely free from sewage and decaying organic matter, and from excess of water. If this is not seen to, then expect typhoid fever from foul subsoil air and polluted well water. Moreover, sand is easily polluted, because the polluted matter on the surface percolates freely into sand with the aid of rain water.

On the other hand, in hot countries, sands are objectionable from their heat, unless they are covered with grass. They do not allow the heat to pass through, but radiate it slowly, and the air is hot over them day and night.

A clay soil is a cold soil, and the air over it is always moister than over dry sand, but a clay soil cannot be easily polluted by sewage water like sand. In some cases fever has been observed to stop on passing from gravel to clay.

On the other hand, the Indian experience, in some cases, has shown that fever death-rates are highest in alluvial clay soils and water-logged ground; whilst the general death-rate was highest on porous wet soils; and that porous wet soils possess no advantage in the way of escape from fever if they be deluged with water.

Fig. 1.

Pervious beds, such as sand and gravel interlaced with impervious beds, such as clay or shale, have the great disadvantage of sweating out water at the outcrop; this is a frequent cause of fever.

A wet hill slope should always on this account be avoided, if at all practicable, but if it must be used, then it must be efficiently drained.

The following instance will explain this. Figure 2 shows the slope of the ground falling towards the plain of Balaclava.

The formations are rock below and above, traversed by a belt of clay and shale.

The 79th Highlanders were placed on the clay, and as the

material was soft, their huts were placed on terraces cut out of the hill-side, and were thus embedded in the ground, and the

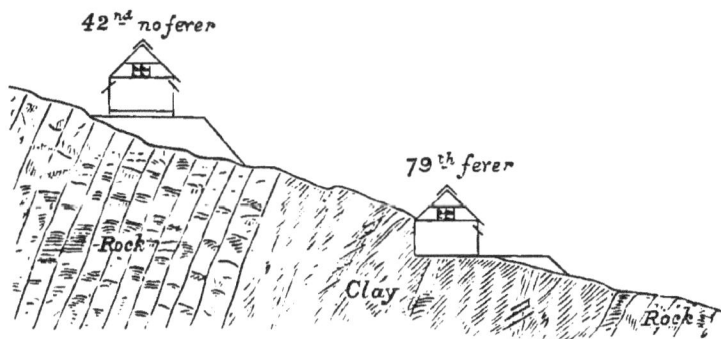

Fig. 2. Huts on hill-side at Balaclava.

floors consequently were always damp. There was no roof ventilation. This regiment had half the men down with fever.

The 42nd Highlanders were placed on the rock, and as it was hard they did not cut into the rock, but preferred building their huts on projecting terraces, so that they were quite dry, and air circulated freely round. This regiment did not suffer from fever.

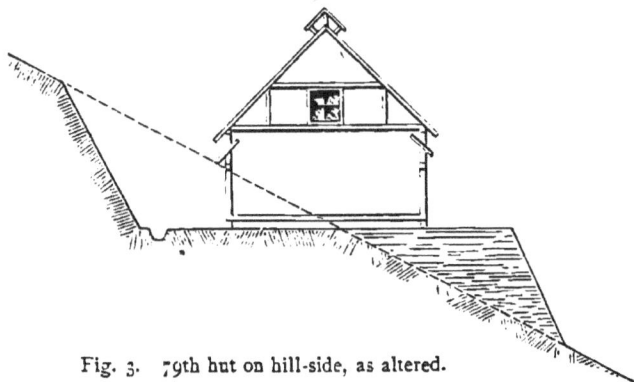

Fig. 3. 79th hut on hill-side, as altered.

The huts on the clay were subsequently altered (Fig. 3), so as to allow of a clear circulation of air. Drainage and roof ventilation were provided, as shown in the sketch, and fever no longer prevailed.

In connection with this, it should be mentioned that where, from circumstances, tents and buildings must be placed on the side of a hill, a plateau should be formed to receive them, and a broad space be left between the hill and the tents or buildings. A trench should also be cut to carry off the moisture between the tent or building and plateau, and no accumulation of refuse should be allowed between the hill and the tent or building.

Although the general features of a site may be unhealthy, when it is absolutely necessary for any reasons, military, political, or otherwise, that it should be occupied, much may be done to remedy its unhealthiness.

For instance, if temporary occupation only is contemplated, probably cutting off the water which may flow from higher levels, or the adoption of measures already mentioned, such as digging trenches round tents or huts, would be all that could be done : but if the ground is to be permanently occupied, not only the area to be built on, but an area extending to probably 100 yards round it on every side should be thoroughly under-drained, and the mouths of the drains so arranged as to allow the aëration of the soil, as well as the removal of the subsoil water. Dwellings should be raised above the level of the ground, and provided with ventilated air spaces underneath.

5. *Effect of the Conditions of the Adjacent Districts on Healthiness of a Site.*

Elevated positions are generally healthy, but when these positions are exposed to wind blowing over marshes or malarial ground, their very elevation is a source of danger.

In order to provide a healthy station at Jamaica, an elevated site from 3,500 to 4,000 feet above the sea level, at a place called New Castle, was selected for barracks. It was situated on the crest of a spur of land falling rapidly from the

Blue Mountains southwards towards the deep damp valleys and ravines, filled with tropical vegetation, which connect the range with the lower country. The sides of the ridge sloped down at angles of 40° and 50°. The surface was clay mixed with vegetable matter. The ridge was so narrow that the huts were placed on terraces cut out of the slopes of the hill, with but a few feet of space between the back of the hut and the soil supporting the terrace above. Even in temperate climates such a position contributes to fever. The result was that in the yellow fever epidemics in 1856 and 1867, those huts which were so placed that the malaria blowing up the valleys must necessarily strike them, yielded a large percentage of yellow fever even at this high elevation.

Algeria perhaps offers some of the best illustrations of the manner in which engineering operations have remedied the evils of the proximity of marshes.

Boná stands on a hill overlooking the sea; a plain of a deep rich vegetable soil extends southward from it, but little raised above the sea level. The plain receives not only the rainfall which falls on its surface, but the water from adjacent mountains, and is consequently saturated with wet. The population living on it and near it suffered intensely from fever ; entire regiments were destroyed by death and inefficiency. It was at last determined to drain the plain. The result of this work was an immediate reduction of the sick and death-rate.

Such instances might be multiplied.

Irrigation, if applied in excess, is unhealthy, and should not be allowed near habitations. In Northern Italy irrigated rice grounds are not allowed within 1000 yards of small towns, and are required to be placed further from large cities ; but in these the water is allowed to stagnate in the subsoil. Where the water is not allowed to stagnate irrigation may be carried on with less danger.

Fondouc, in Algeria, is situated on sloping ground, imme-

diately above the marshy plain of the Mitidja, and has
mountain ranges behind it. It was first occupied in 1844,
and in the succeeding year half the population was swept
away by fevers and dysentery. During the first 20 years the
mortality was 10 per cent. The surrounding marsh has been
cultivated, and there are now upwards of 10 square miles
round the town under cultivation, producing cereals, cotton,
tobacco, and wine.

The cultivation consists of ploughing and trenching combined with irrigation, so that water in excess is not applied.
The mortality now is only 20 per 1000.

In India, wherever water is applied in excess for irrigation,
so as to become stagnated in the subsoil, ague, spleen disease and fever prevail.

But it is possible to improve such localities by draining
away the superfluous stagnant subsoil water.

In the northern Doab districts in the North-West Provinces
of India a great diminution of the excessive fever mortality
for which these districts were noted has followed the extension
of drainage works, by which the water which formerly stagnated on and in the land is now led away by continuously
flowing streams.

But it is not sufficient to make the drains. All drainage
cuts are liable to become injured, if open, by vegetation, and
in all cases by decay, by atmospheric causes and other means.
If not properly maintained and cleared out, the evils they are
created to remove will recur.

For instance, at Boná, from which, as already mentioned, in
consequence of drainage works the fever disappeared.

The drains were left to atmospheric influences; they became partially obstructed and irregular, and did not allow the
water to reach the outfall; the result was a violent outbreak
of fever at Boná attended with great loss of life both civil and
military, an enquiry took place, the drainage was rectified,
and since then Boná has been healthy.

The engineer thus has it in his hands to mitigate the evils of a marshy district by providing for the removal of stagnant water, and to prevent the evils which arise from irrigation by combining drainage with works intended for irrigation.

6. *Summary of Conclusions as to a Healthy Site.*

The following is a brief summary of the conclusions to which the considerations above adduced point—

1. Clay soils should, if possible, be avoided.

2. Ground at the foot of a slope, or in deep valleys, which receives drainage from higher levels, should be avoided. It predisposes its occupants, even in temperate climates, to epidemic diseases.

3. High positions exposed to winds blowing over low marshy ground, although miles away, are in certain climates unsafe, on account of fevers. Indeed, it sometimes happens that a site in the immediate vicinity of a marsh, or other local cause of disease, especially if protected by a screen of wood, is safer than an elevated and distant position to leeward.

4. Elevated sites situated on the margin or at the heads of steep ravines, up which malaria may be carried by air currents flowing upwards from the low country, are apt to become unhealthy at particular seasons. Such ravines, moreover, from want of care, are often made receptacles for decaying matter and filth, and become dangerous nuisances. In tropical climates these ravines convey malaria, and occasion aggravated remittent, or even yellow fevers, at an elevation which would be otherwise exempt from the action of tropical malaria.

5. Ground covered with rank vegetation, especially in tropical climates, is unhealthy, partly on account of the amount of decaying matter in the soil, partly because the presence of such vegetation is in itself a mark of the presence of subsoil water, or of a humid atmosphere.

6. In warm climates, muddy sea beaches, or river banks, or muddy ground generally, if it be subject to periodical flooding, and marsh land, especially if it be partly covered with mixed salt and fresh water, are peculiarly hazardous to health.

7. A porous subsoil, not encumbered with vegetation, with a good fall for drainage, not receiving or retaining the water from any higher ground, and the prevailing winds blowing over no marshy or unwholesome ground, will, as a general rule, afford the greatest amount of protection from disease which the climate admits of.

8. To test the healthiness of a site an enquiry into the rate of sickness and mortality in the district will afford valuable information. But care should be taken not to be guided by the mortality alone. The nature of the diseases, and the facility, or otherwise, with which convalescences and recoveries take place, must also be taken into account.

To sum up these conclusions for this country for a site to be selected for occupation. The local climate should be healthy; the soil should be dry and porous; it should be protected from the north and east by shelter at a sufficient distance to prevent stagnation of air or damp,—otherwise the shelter from cold and unhealthy winds, which is an evil recurring only at intervals, will be purchased by loss of healthiness at other times. The ground should fall in all directions, to facilitate drainage ; it should not be on a steep slope, for high ground rising near a building stagnates the air just as a wall stagnates it : the natural drainage outlets should be sufficient and available. There should be nothing to prevent a perfectly free circulation of air over the district ; there should be no nuisances, damp ravines, muddy creeks or ditches, undrained or marshy ground close to the site, or in such a position that the prevailing winds would blow the effluvia over it.

The site should, moreover, be thoroughly under-drained,

except possibly in a case where the ground is so elevated and porous as to ensure that water never remains in it ; and if there is higher ground adjacent, the water from the higher ground should be carefully cut off by underground catch-water drains, and led away from the vicinity of the site.

The object to be attained in laying out the ground is the rapid and effectual removal of all water from the buildings themselves, and from the ground in their vicinity, so that there shall be no stagnation in or near the site.

It is no doubt impossible always to procure a perfect site for building ; but it will be necessary in the construction of buildings upon a given site to discount any departure from these qualifications by additional sanitary precautions in the building—i. e. by increased expenditure.

CHAPTER IV.

HOWEVER healthy a site may be, evil may accrue from the undue crowding of buildings upon it, and from the disregard of other conditions necessary for securing purity of air where numbers of persons are congregated.

The deteriorating effect of residence in towns has been frequently noticed. The Registrar-General has shown that a population of 12,892,982 persons living on 3,183,965 acres in the districts comprising the chief towns of England, showed an average death-rate for ten years of 24.4 per 1000 ; whilst a population of 9,819.284 living on 34,135.256 acres in districts comprising small towns and country parishes, showed an average death-rate for a similar period of 19.4 per 1000.

Dr. Morgan's paper on the deterioration of races in great cities shows that of the adult population of London 53 per cent., of that of Birmingham 49 per cent., of that of Manchester 50 per cent., and of that of Liverpool 62 per cent., were immigrants from the country settled in the town, and that the majority of the incomers were men and women in the prime of life.

The mortality in these four towns averaged 26 per 1000 against 19 per 1000 in the adjacent country districts ; the mortality of persons under the age of 15 being 40.7 per 1000 in these towns against 22 per 1000 in the country districts.

The marriages in the city population were four times as numerous as in the agricultural counties, but the births in the town population only exceeded those in the agricultural population by one-sixth.

A statistical analysis by Mr. Francis Galton of the details of 1000 town families and 1000 country families, selected from the town of Coventry and the adjacent agricultural population, showed that the town population supplied to the next generation only three-quarters of the number of adults supplied by the equally numerous country population; and that in two generations the adult grandchildren of artisan townsfolk were little more than half as numerous as those of labouring people who lived in healthy country districts. In large, closely-built centres of population the ratio would probably be considerably increased against the town population.

The greater unhealthiness of towns is largely due to the too close proximity of the dwellings, the consequent absence of fresh air, and the saturation of the subsoil with impurities passing into it from the closely occupied surface.

The subsoil of all large Indian cities has been saturated with the filth of generations, just as the subsoil of large cities in ancient times had been saturated.

From this saturation these ancient cities became foci of disease, and were abandoned.

Many large cities in India contain saltpetre factories, the nitrogen being derived from the previous contamination of the soil with decaying animal matter.

The careful paving of a town area, coupled with adequate drainage to carry off the water falling on the gravel surface, is a great protection to health.

It is essential for health that buildings should have free circulation of air all round them, and as much sunlight as possible.

Sunshine and direct daylight from the sky purify air and

promote vitality. Sunshine kills certain classes of bacteria and their spores, e.g. anthrax, and is antagonistic to some other micro-organisms connected with disease.

In a street adequate sunlight on the face of a house will not be obtained in this country unless the width of the street is equal to twice the height of the buildings which are interposed between the face of the house and the sun.

Rows of back-to-back dwellings which do not admit of thorough ventilation should not be permitted; there should be a clear space sufficient for a free circulation of air at the back of every dwelling house, from the level of the lowest floor or basement, and along its whole width between it and the building behind it. Corner houses of streets should not be shut in by houses close to them on both sides, without any opening at the back; they should have a courtyard on one side or the other. Dead ends of courts or streets should not be permitted.

In temperate and cold climates buildings should, if possible, be arranged so as to allow the sun to shine on both of the principal sides of the building during the course of the day.

Indeed in the general arrangement of buildings or tents or huts, it may be laid down as a safe rule that the larger the space that is allowed between them for circulation of air the more healthy will the occupants be. Hence, if for any reason it be necessary to lodge a large number of persons on a given limited area, it is preferable for health to place them in buildings of several storeys high, designed to afford free through ventilation, and so distributed on the site as to admit of free circulation of air all round each building, rather than in dwellings of lesser height which must be in close proximity, not admitting of free circulation of air. But at the same time care must be taken in lofty buildings that there shall be no community of air between the several storeys; in buildings where this has been permitted, the health is deteriorated.

The following are examples of the death-rates in some of the most densely peopled districts in the metropolis, as compared with less dense localities :—

	Approximate Number of inhabitants per acre.	Number of Houses per acre.	Death-rate.
St. Anne's, Soho	331·35	25·24 ⎫	24·16 [1]
Strand	241·05	22·23 ⎭	
Brompton	41·05	5·81	22·4 [2]
Lewisham	3·05	0·56 ⎫	18·8 [3]
Eltham	1·04	0·18 ⎭	
Metropolis	43	5·55	22·3
Model Lodging Houses, viz. . ⎫ Industrial Lodgings Company [4] ⎭	860		16
Metropolitan Association	1140	...	18

The examples drawn from towns are from places where paving and draining have been more or less carried out, and where, nevertheless, the influence of surface overcrowding on health is obvious on a comparison being made with less crowded districts.

The superior healthiness of the Model Lodging Houses is due partly to the careful provision of sanitary arrangements, but mainly to the fact that the numerous storeys in these buildings, whilst affording accommodation for a dense population on a limited area, are provided with free through ventilation, and circulation of foul air from lower to upper storeys is prevented; in addition to this, ample space is provided all round the structure for the circulation of air, and impurities

[1] Registrar's District, Strand.
[2]   „   „   Kensington.
[3]   „   „   Lewisham.
[4] The population per acre in one of the buildings of this Company is 2,200.

D

are not allowed to be retained on the open area round the buildings.

With armies, a camp is formed of a number of tents or huts with circulation of air all round each. But if the force be large, a too close proximity of tents may be, and certainly has been, a common cause of camp diseases. A camp is a temporary town, without paving or proper drainage. It is only by paving and drainage that the deleterious influence of surface overcrowding in towns can be reduced to a minimum. But paving and drainage cannot be carried out to a sufficient extent in camps to enable the surface to be crowded with safety to health ; and therefore in laying out a camp the extent of space allotted to it should be as large as the nature of the ground, or of the service, will admit ; great care should be exercised in the selection of places for the deposit of refuse, and after a temporary camp has been occupied for some time, the site should be abandoned for a new one.

The Quartermaster-General's instructions for camping, issued at the commencement of the Crimean War, authorised densities of population on the camp surface equal to 542 and 1037 inhabitants per acre. The lowest of these densities is nearly double that of St. Anne's, Soho, one of the most densely populated districts in England, where the population occupies houses of ordinary construction. It includes not only the ground actually covered by tents, but all the open spaces in the camp. The ground actually covered by tents in these plans of encampment gave a density of population equal to 1632 per acre, or a space of little over 5 × 5 feet for each individual.

A comparison of these authorised densities for camps, which had neither drainage nor paving, with the dense populations in towns already mentioned, affords an index of what would be likely to be the influence on health of surface overcrowding in camps where the occupation of the same ground is protracted.

The surface area per tent for different densities of population per square mile is as follows :—

Square yards per tent.	Tents per acre.	Troops per acre, assuming 12 men per tent.
50	96·6	1159·2
100	48·4	580·8
400	12·1	145·2
1000	4·84	58·

The number of troops to be placed on a given area must be determined by local circumstances, but the above table will be useful in enabling a correct judgment to be formed upon one very important element in the sanitary state of camps—namely, density of population.

The manner of arranging tents is of importance to health, as well as to cleanliness.

Battalion tents should never be arranged in double line; short single lines are best. The tents in line should be separated from each other by a space at the very least equal to a diameter and a half of a tent, and the farther the lines can be conveniently placed from each other the better. These are all matters which are necessarily more or less subject to military considerations, and therefore the object in pointing them out is to furnish a sanitary standard at which to aim, rather than to suggest an absolute rule.

The construction of permanent buildings is a matter over which the architect and the engineer have a more complete control than over the location and arrangement of a camp.

The general position of houses, hospitals, asylums, prisons, or barracks, and even towns, is no doubt settled by the special circumstances of each case; or, in the case of government buildings, by military or political considerations; but the actual site within such general locality, and the arrangement of buildings on the site, is a matter which falls more immediately within the province of the architect or engineer to determine.

The health of any building is dependent upon free-moving pure air, outside and inside its walls; anything which interferes with this first condition of health is injurious.

If the building is placed in a town, the health of the inmates is governed by the same conditions as those of the rest of the population of the town.

Thus certain barracks in manufacturing towns in 1875-80 showed a death-rate of 10·88 per 1000, as compared with a death-rate of 6·98 per 1000 at Aldershot. But the continued occupation of Aldershot as a camp, on a porous soil, without paving or adequate drainage round the huts, led to a gradual deterioration of the health of the camp; and brick buildings with drainage and a protected surface are being substituted for the temporary wooden huts.

Where commercial, political, military, or other necessities require that a building containing a large number of inmates should be placed in a town, additional precautions must be taken to render it as little unhealthy as possible. The enclosure should be sufficient to allow of ample space being reserved for a supply of fresh air between the enclosure wall and the inhabited buildings, and between the buildings themselves.

The sources of impure air within the enclosure, such as ash-pits, manure-pits, &c., should be reduced to a minimum, and so placed that the air in their vicinity shall not stagnate.

In the design of any building intended for habitation the first consideration is, how can the ground at the disposal of the architect or engineer be best utilised, so as to secure pure flowing air and sunlight over every part of the building?

CHAPTER V.

HAVING explained the conditions which govern the purity of the outer air, it is in the next place necessary to consider what is the accepted meaning of purity of air in an inhabited building.

In order to appreciate the enormous difference between the purity of air out of doors and the purity of air in a confined space, it is necessary to consider what are the constituents of the outer air.

Standard of Purity in Air.

Air taken in the country under the most favourable circumstances, in free open spaces or on elevated ground, consists of the following constituents (Angus Smith) :—

1000 parts [1] { Oxygen, 209 to 211 parts ; 209·6 mean.
{ Nitrogen, 789 to 791 parts.

Moreover, every analysis of air shows the presence in varying proportions of carbonic acid, vapour of water, organic matter, ammonia, suspended matter.

The purity or impurity of the air, and its effect on health, depends upon the greater or less degree in which these various subsidiary matters are present in the air.

The most important of the gaseous impurities in air which have an influence on health is carbonic acid CO_2. It is,

[1] The newly discovered constituent, e. g. Argon, may be neglected for this argument.

however, less important on account of its own special action, than because of its use as a measure of the purity of the air.

The average amount of CO_2 has been taken at 0·400 volumes per 1000 in normal air, although it is not unfrequently as low as ·2, and sometimes as high as ·5, or more.

M. Reiset obtained from a year's observation, at a station in the country far from dwellings, and situated at about four miles from Dieppe, an average of ·2942 per 1000. The air above a crop of red trefoil in the month of June gave ·2898 ; and at a height of one foot from the soil, in a barley-field in July, ·2829, per 1000: the corresponding amounts at the country station being ·2915 and ·2933 per 1000 respectively. The presence of 300 sheep near the apparatus raised the proportion to ·3178 per 1000, and at Paris in May 1873–75–79 the mean amount was ·3027 per 1000.

The presence of from 1·5 to 2 per cent. of this gas produces in many persons severe headache; but as much as 3 per cent., or even more, has been found just endurable under certain circumstances. A candle will be extinguished with 2·5 per cent.; and it may be assumed that the presence of 5 per cent. or over will cause death.

In this connection, it may be mentioned that carbonic oxyde CO_1 is eminently poisonous. Less than 0·5 per cent. has produced poisonous symptoms, and 1 per cent. rapidly produces fatal results. This gas is formed by the imperfect combustion of carbon.

The deleterious action of gases disengaged from marshes has been attributed to the presence of sulphuretted hydrogen, but the action of these gases is not very accurately determined.

There are, moreover, various suspended matters in air which produce disease from mechanical causes, such as the dust which in Egypt produces a sort of ophthalmia. Bronchitis and lung disease prevail in many factories, arising from the inhalations by the workmen of the dust of coal, sand, and steel, or of particles of cotton or hemp. Stonemasons suffer from inhalation of stone-dust.

The Guards suffered largely about 40 years ago from lung disease ; one of the contributing causes was assumed to be the quantity of pipeclay they inhaled in the process of cleaning their white cloth fatigue jackets.

House-painters suffer from the dust of white lead ; though in this, as in many cases, the persons suffer as much from swallowing particles, in consequence of not washing off the dirt from their hands before eating, as from breathing the dust.

Of all the impurities of air, that which stands highest in the scale of injury to health is organic matter. An undue proportion of carbonic acid may indeed kill outright, but to the results of the presence of organic matter diseases of impure air are mainly traceable.

Malaria appears to arise from the poison of decaying moist vegetable matter in marshes and forests. Typhoid fever is traceable to the poisonous air which arises from putrefactive substances. In camps men have had typhoid fever in consequence of their tents being placed near to manure heaps, or on damp soil. Phthisis and other diseases may result from breathing air rendered impure by the putrefying organic matter thrown off from the human body in the process of breathing and transpiring.

So long as air is in movement out of doors, the products of vegetable and animal waste are being continually removed by the air from the vicinity of their origin. They are washed out of the air by rain. or removed by snow and hail. They are removed by oxidation, much of which is probably due to the action of ozone, and would not be effected by ordinary oxygen.

Ozone is oxygen in an altered or allotropic condition, and appears to be formed by the passage of the electric spark through dry oxygen or by slow oxidation of phosphorus and other essential oils in presence of moisture. Ozone is insoluble in water.

Ozone is rarely, if ever, absent in fine weather from the air of the country; but it is more abundant, on the whole, in the air of the mountain than of the plain. It is also said to occur in larger quantity near to the sea than in inland districts. It has been found to an unusual amount after thunderstorms.

There is great variety of opinion as to the conditions which produce ozone. According to some observers, the amount of ozone in the air is greater in winter than in summer, and greater in spring than in autumn; but according to other observers, it is greater in spring and summer than in autumn and winter. Ozone has usually been found more abundantly in the air at night than by day; but, again, some careful observers have found the reverse of this statement to be true.

No connection has yet been proved to exist between the amount of ozone in the atmosphere and the occurrence of epidemic and other forms of disease.

Ozone is rarely found in the air of large towns, unless in a suburb when the wind is blowing from the country; and it is only under the rarest and most exceptional conditions that it is found in the air of the largest and best ventilated apartments. It is, in fact, rapidly destroyed by smoke and other impurities which are present in the air of localities where large bodies of men have fixed their habitations.

The permanent absence of ozone from the air of a locality may, however, be regarded as a proof that the air is adulterated air. Its absence from the air of towns and of large rooms, even in the country, is probably the chief cause of the difference which every one feels when he breathes the air of a town or of an apartment, however spacious, and afterwards inhales the fresh or ozone-containing air of the open country.

The amount of ozone in the atmosphere is extremely small, and an excess of ozone is destructive to life; thus the respiration for a very short time of oxygen containing about 1-240th part of ozone is certainly fatal to all animals; whilst they would live in good health for months after respiring oxygen alone for 37 hours, the carbonic acid being removed during the experiment.

The action of breathing and transpiring upon the air is as follows :—

1. The oxygen is diminished.
2. The carbonic acid is increased.
3. A large amount of watery vapour is added.
4. There is an evolution of ammonia and organic matter.
5. A considerable amount of suspended matter is set free, consisting of epithelium, and molecular and cellular matter, in a more or less active condition of putrefaction. At the same time, portions of epithelium are constantly being given off from the skin, and even pus cells from suppurating surfaces ; as, for instance, with surgical cases in hospitals.

The oxygen is, of course, diminished in the direct ratio of the consumption of carbon and hydrogen in the system. As regards the amount of carbonic acid, a subsistence diet, sufficient for the internal work of the body only, is a little under 3000 grains of carbon daily, yielding about 13·6 cubic feet, or about 0·57 cubic feet per hour of CO_2.

Angus Smith, in his experiments, was unable to find more than 0·4 per hour of CO_2 given off; but the experiments of Pettenkofer showed that in a state of repose an adult gave off about 0·7, and in a state of active work 0·9 to 1·0 or more. The constitution and usual diet of the person experimented on no doubt influences the result.

These numbers correspond pretty closely with theoretical calculation, but if the number 0·6 be taken to allow for difference of age, weight, and sex, it will be well within the mark in the calculation.

The amount of vapour varies, but taking the amount from skin and lungs together, it may be assumed at about 30 oz. per diem, or about 550 grains per hour, enough to saturate about 90 cubic feet of air at a temperature of 63° Fahrenheit.

The amount of organic matter has been variously estimated, but there are hardly any trustworthy experiments on record.

As already mentioned, this organic matter is highly poisonous; and it is as much from the presence of this as from carbonic acid in re-breathed air that injury arises.

The air of towns is rendered impure chiefly from the presence of suspended matters.

The experiments of Dr. Angus Smith show that in towns the oxygen is not less than in the country districts, and that carbonic acid is not materially in excess.

See the following table :—

In Manchester.

Oxygen.		CO_2	
	Per 1000		Per 1000 vols.
In fog and frost	209·100	Streets	0·403
Outer circle, not raining .	209·407	Where fields begin . . .	0·369
Suburb, in wet weather .	{ 206·800 / 209·600	Streets in fog	0·679

In London.

Oxygen.		CO_2	
Open places, summer . .	209·500	On Thames	0·343
Streets, November . . .	208·850	Parks, open	0·301
		Streets	0·380

It is therefore to other impurities that the oppression from town air is attributable. For instance—The presence of sulphuric acid in the air is very noteworthy.

Numerous analyses of various sorts of coal showed that whilst there was a mean of 1.7 per cent. of sulphur in the several coals, no more than 0·2 per cent. remained in the ash. Therefore the burning of 1000 tons of coals of this description would send 15 tons of sulphur into the air as sulphurous acid ; and this soon becomes converted into sulphuric acid ; this is sufficient in quantity to render the rain water which is collected in towns very frequently acid.

It has been estimated that the coal consumed in Glasgow

and its vicinity gives off sufficient sulphur acids to amount to 300,000 tons of oil of vitriol annually. The quantity of coal estimated to be consumed annually in London is about 5,000,000 tons, which, from this calculation, would send into the air 75.000 tons of sulphurous acid. London air contains about 19 grains of sulphurous acid in a cubic yard of air. It contains, moreover, an enormous quantity of soot, fine carbon and tarry particles of coal ; of the two latter almost 1 per cent. is given off in combustion. This rarely rises above 600 feet from the ground. London air also contains much suspended organic matter, independently of the sewer and other emanations, and independently of the ammonia given out by the manure of the enormous number of horses kept in London. It is noteworthy that the mud from a paved street in London was found on analysis to contain nearly 90 per cent. of horses' dung ; the mud on the new wood pavements consists almost entirely of horse dung[1]. In addition to such matters town air contains vestiges of food, clothing, and building materials ; as well as dust from manufactories.

An important cause of the impurity of air in town houses especially arises from the use of coal-gas in rooms.

The products of the combustion of gas are carbonic acid, carbonic oxide, compounds of ammonia, and various compounds of sulphur, which are injurious to health.

The products of combustion vary much with the quality of the gas and the completeness of the process, but 100 cubic feet will unite with from 90 to 164 cubic feet of oxygen, and produce 200 cubic feet of carbonic acid, and from 20 to 50 grains of sulphuric acid, so that 100 cubic feet of coal-gas consume the oxygen or destroy the vital qualities of 800 cubic feet of air, and raise the temperature of 31·290 cubic feet of air 100° Fahrenheit.

---

[1] The wood pavement laid down in Regent Street more than 30 years ago was removed because it had become so saturated with ammonia that the emanations tarnished the plate in silversmiths' shops.

With imperfect combustion, 67 per cent. of nitrogen, 16 per cent. of water, 7 per cent. of carbonic acid, and 5 to 6 per cent. of carbonic oxide, with sulphurous acid and ammonia, are thrown into the atmosphere, but the quantity of carbonic oxide will be materially reduced with more perfect combustion.

It follows that each cubic foot of gas burnt per hour may be assumed upon an average to vitiate as much air as would be rendered impure by the respiration of an individual. There is the further danger that a gas-burner in a room, even when not in use, may allow of a slight leakage of gas, and thus cause discomfort, if not danger.

An oil lamp burning 154 grains of oil per hour consumes the oxygen of 3.2 cubic feet of air, and produces a little more than ·5 cubic feet of carbonic acid. The combustion does not produce the compounds of sulphur which result from the use of coal gas.

In the open country, the atmospheric currents continually disperse the various substances thrown off in breathing.

The movement of the air is stated in the Registrar General's reports to be about 12 miles an hour, on an average, or rather more than 17 feet per second. It will rarely be much below 6 feet per second.

Imagine a frame about the height and width of a human body, measuring about 6 feet by 1½, or 9 square feet; multiplying this by the velocity of movement of the air at 6 feet a second, it will appear that in one second 54 cubic feet, in one minute 3240 cubic feet, in one hour 196,400 cubic feet, of air would flow over one person in the open.

In a room the conditions are very different. In barracks, in a temperate climate, 600 cubic feet is the space allotted by regulation to each soldier; and when in hospital from 1000 to 2500 cubic feet to each patient

If it were desired to supply in a room a volume of fresh air comparable with that supplied out of doors, it would be necessary to change the air of the room from twice to six times in every minute, but this would be a practical impossibility; and even if it were possible, it would entail conditions very disagreeable to the occupants.

It is thus evident that when considering the condition of air indoors, it is necessary to seek a standard of admissible impurity in the air, rather than a standard of purity of air, comparable with that which exists out of doors.

In judging of the amount of impurity which may be allowed in an inhabited air-space, the sense of smell, when carefully educated, affords the best indication of the relative purity and impurity of different kinds of air.

The accompanying table obtained from results of experiments communicated by Dr. de Chaumont to the Royal Society shows the conclusions at which he arrived from a very large number of observations on the air of barracks and hospitals. The method employed in judging of the quality of the air was to enter directly from the open air into the room in which the air was to be judged, after having been at least 15 minutes in the open air. It will be seen how closely the state of the room, as detected by the sense of smell, agrees with that which would be expected from the carbonic acid as shown by analysis.

Sense of Smell.	Temperature.		Vapour.		Carbonic Acid per 1000 volumes.	
	In air space.	Excess over outer air.	In air space.	Excess over outer air.	In room.	Excess over outer air.
Fresh	62 85	5.38	4.629	0.344	0.5999	0.1830
A little smell	62.85	8.00	4.823	0.687	0.8004	0.3894
Close or disagreeable smell .	64.67	12.91	4.909	1.072	1.0027	0.6322
Very close, or offensive and oppressive smell . . .	65.15	13.87	5.078	1.409	1.2335	0.8432
Extremely close, when the sense of smell can no longer differentiate	65.05	13.19	5.194	1.319	1.2818	0.8817

In these experiments, Dr. de Chaumont takes ·0002 of

carbonic acid per cubic foot as the standard of impurity, in addition to ·0004 of carbonic acid per cubic foot as the normal amount of CO_2 in the outer air.

The experiments were made in barracks and in hospitals, and a result came out from the experiments confirmatory of the opinion that, in the case of sick men, more air is required to keep the air space pure to the senses than is necessary in the case of men in health. It appeared that, in barracks, the mean amount of respiratory carbonic acid, when the air was pure to the senses, was ·196 per 1000 volumes, but in hospitals it was only ·157 ; or, in other words, whilst in the hospitals the air would have smelt somewhat impure when the CO_2 was ·157, in the barracks with that amount, it was fresh.

On these grounds it would therefore appear that whilst the standard for impurity for healthy persons may be regulated by allowing an excess of ·0002 per cubic foot of CO_2 over that in the outer air, it would be desirable to limit the excess in the case of sick to ·00015 per cubic foot.

In addition to the proportion of carbonic acid, and of the impurities of which its presence affords a rough test, there are conditions of temperature and humidity necessary for good ventilation.

Temperature. The dry bulb thermometer in this climate ought to read 63° F. to 65° F., and ought not, if possible, to fall much below 60° F.

The wet bulb ought to read 58° F. to 61° F. That is to say, in this country the difference between the two thermometers ought not to be less than 4° F. or more than 8° F. A greater degree of dryness in the air, provided the supply of air be ample, is not however found objectionable.

In the open air, in healthy weather, it is often 8° or 9° or more. The difference is of course increased in hot and dry climates.

Vapour ought not to exceed 4·7 grains per cubic foot at a temperature of 63°, or 5·0 grains at a temperature of 65° F.

The limit of humidity is 75 per cent. or under.

When the outer air is saturated, as in wet weather, the reduction of the humidity in a room will depend on the increase of temperature of the air admitted.

The capacity of the air for moisture increases enormously with the temperature, and that which would saturate air at 50° F. would give only 71 per cent. at 60° F. Thus, at 50° F. a cubic foot of air is saturated by 4·1 grains; but at 60° F. it requires 5·8 grains, so that 4·1 grains would give only 71 per cent.

If therefore the outer air is at a temperature of 50°, and if the temperature inside the room be maintained at a comfortable standard, say 63° to 65°, the incoming moisture would never cause an excess of humidity.

In the case of an external atmosphere, saturated at or above the temperature within, such as occurs occasionally in hot climates, it would be necessary to let in an unlimited quantity of air through every possible aperture.

CHAPTER VI.

THE purity of the air within an inhabited space, enclosed on all sides, is necessarily vitiated by the emanations proceeding from the bodies of those who inhabit it, and especially by the effect on it of their respirations. With persons suffering from disease, especially infectious fevers, or from wounds, or sores, these emanations are greater in quantity and more poisonous in quality, than from persons in health. Stagnation in the movement of the air would lead to rapid putrefaction of these emanations.

Vitiated air does not necessarily mix with the whole air of the room with rapidity. Any one may satisfy himself of this by comparing the upper part of a heated room with the lower, or by examining outlets for the escape of air. The top of a room will sometimes be found to contain much more carbonic acid than the lower part. Therefore under certain conditions the law of diffusion does not act so rapidly as to prevent an occasional difference between the amount of CO_2 in the upper and the lower air of a room ; the organic emanations diffuse themselves more slowly, and without absolute uniformity, and are deposited on the cool walls, ceilings, floor, and furniture, where they may be easily collected if desired.

It would be desirable, if it were practicable, to remove the exhalations with such rapidity as to prevent deposition, and to permit as little admixture as possible with the air of the room ; but this is not practicable.

In considering theoretically the condition of a room in which sources of impurity exist, and which is furnished with any kind of ventilating arrangements, the two *extreme* suppositions (both inadmissible) are

1. That all exhalations are *immediately* removed completely out of contact of the persons in it, so that the occupants of the room are in the same condition as to purity of air as if they were out of doors in a brisk wind.

2. That the ventilation is so unequal that the spaces immediately surrounding the persons do not get ventilated at all ; and that the occupants of the room practically live in an air which may become saturated with noxious matter.

The actual state of things must be something between these two. And it is probable that the *best* condition actually attainable would approximate, not very closely, but still in some tolerable degree, to the ideal condition in which all diffusible emanations should be *instantaneously* and uniformly diffused through the whole space.

Now, supposing this ideal condition to subsist, it is perfectly easy to show that the degree of purity of the air would ultimately depend in no way on the size of the room, but solely on these two things, viz. (a) the rate at which emanations are produced : (β) the rate at which fresh air is admitted.

Demonstration [1].

Suppose P units of diffusible poison are produced per hour.

Also suppose A cubic feet of air introduced per hour.

The same number must necessarily escape per hour.

The condition of the room having become permanent, the quantity of poison escaping is the same per hour as the quantity produced (otherwise the condition of the room would be changing).

---

[1] By the late Professor Donkin of Oxford, in Report of Royal Commission on Cubic Space in Workhouses.

Hence P units of poison escape per hour; and since this quantity is carried away in A cubic feet of air, the escaping air necessarily contains $\dfrac{P}{A}$ units of poison per cubic foot, whatever be the size of the room. The escaping air may or may not be a sample of the average of the room. On the supposition of uniform diffusion, *it is a sample*; hence :—

On the supposition of uniform diffusion, the air in the room ultimately contains $\dfrac{P}{A}$ units of poison per cubic foot, whatever be its size. Thus, on this supposition, the final condition of the air depends only on the rate of production of poison, and on the rate of admission of fresh air, and in no way on the space.

But if the mode of ventilation be bad, the diffusion will not be uniform; and not only so, but there is theoretically no limit (except that of saturation) to the quantity of poison which may remain as a constant quantity in the room, however abundant may be the supply of fresh air.

It seems hardly conceivable, though it is mathematically possible, that the whole quantity of poison remaining permanently in the room could be reduced by any contrivance *below* that of uniform diffusion.

The condition referred to above, as *permanent*, is a state which, theoretically, would never be actually attained, but to which the actual condition would continually approximate as a limit.

Suppose, as before, that P units of poison are produced in the room per hour when it is occupied. Suppose also that the fresh air itself contains p units of poison per cubic foot. Let c be the number of cubic feet in the room; and suppose that at a given time the room begins to be occupied, and that A cubic feet per hour of fresh air are introduced, so that the same volume of air per hour also escapes.

Then, if x be the number of units of poison per cubic foot in the air of the room at the end of t hours, it can be shown (see demonstration below [1]) that on the hypothesis of uniform diffusion

$$x = p + \frac{P}{A} - \frac{P}{A}\epsilon^{-\frac{At}{c}},$$

where ϵ is (as usual) the number 2·718. The numerical value of the last term in this expression diminishes rapidly as t increases, and will become insensible after a number of hours depending on the ratio of A to c. Thus the quantity of poison per cubic foot increases continually from the initial

---

[1] *Demonstration by the late Professor Donkin of Oxford of the formula used above.*

C = content of room in cubic feet.
A = number of cubic feet of fresh air introduced per hour.
P = number of units of poison produced in room per hour.
p = number of units of poison in a cubic foot of fresh air.
t = time (in hours) since beginning of occupation.
x = number of units of poison per cubic foot in the room at time t.

During the next instant dt, Adt cubic feet of air are introduced, and the same quantity escapes.

The escaping air contains x units of poison per foot, so that $Axdt$ is the quantity of poison which escapes.

During the same instant, $Apdt$ units are introduced with the fresh air, and Pdt units are produced in the room. Hence the whole increase of poison in the room is

$$(P + Ap - Ax)\,dt,$$

but the increase per cubic foot is dx, so that the whole increase is cdx; hence

$$cdx = (P + Ap - Ax)\,dt.$$

Integrating this equation, and determining the constant of integration by the condition that $x = p$ when $t = 0$, we obtain the expression for x given above.

The following may be added.

Suppose a room of c cubic feet, containing initially π' units of poison per cubic foot, to be shut up for t hours with a man in it who produces P units per hour. At the end of that time, how much fresh air (containing p units per cubic foot) must be added to the air of the room in order to reduce the quantity of poison per cubic foot to π units?

It is easily found that the number of cubic feet required is

$$\frac{c(\pi' - \pi) + Pt}{\pi - p};$$

and this formula shows clearly that if $\pi' = \pi$, that is, if the room is to be brought back to its initial condition, the quantity required is independent of c, that is, of the size of the room.

E 2

value p, and tends to the final or permanent value $p+\dfrac{P}{A}$, which it will attain *sensibly* after a finite number of hours, though never rigorously.

The size of the room then does not affect the permanent condition of the air; but everything else being the same, the larger the room is, the longer it will be (after beginning to be occupied) before it attains sensibly its final or permanent condition of impurity.

If, in the final condition, the number of units of poison per cubic foot be π, then $\pi = p + \dfrac{P}{A}$,

$$\text{whence} \qquad A = \frac{P}{\pi - p},$$

which gives the number of cubic feet of fresh air per hour required to maintain this condition.

For example ; suppose a man produces 6 units of carbonic acid per hour, and fresh air contains ·004 such units per cubic foot, if it is required to maintain a room (of whatever size), constantly occupied by one man, in such a condition that the units of carbonic acid in a cubic foot shall never exceed ·006,

$$\text{then} \qquad A = \frac{6}{·006 - ·004} = 3000,$$

that is, 3000 cubic feet of fresh air must be supplied per hour.

In this case, at the end of t hours after the room begins to be occupied, the number of units of carbonic acid per cubic foot is

$$·006 - ·002 \times \epsilon^{-\frac{3,000\,t}{c}},$$

where c is the number of cubic feet of space in the room.

Thus, suppose the room contains 1000 cubic feet of space, then the units of carbonic acid per cubic foot are

at first	. ·004,
after 1 hour	. ·005900,
„ 2 hours .	. . ·005995,
„ 3 „ .	. ·0059997 ;

so that after two hours the room would have *sensibly* reached the final condition of ·006 units per cubic foot. If the room contained only 100 cubic feet of space, the approximation to the final state would be much more rapid.

In considering the question of purity of air in an enclosed space, it is necessary to take into consideration the sources of vapour inside the room. Every man gives off from lungs and skin each hour enough to raise the humidity from 70 per cent. to complete saturation in 500 cubic feet at 60° F., and to raise it to 82 per cent. in 1500 cubic feet. Now to reduce this amount to 73 per cent. would take 3000 cubic feet of air saturated at 50° F. But the vapour given off by the body is not the only source of humidity. Humidity may arise from the combustion of lights, or the vapour of liquids used in the room.

According to this theoretical assumption of temperature and moisture, a room containing an air space of 1000 cubic feet, occupied by one individual, would require to be supplied with 3000 cubic feet per hour, in order to maintain it in a proper condition of purity and humidity.

Thus, upon the assumption made, the theoretical calculations, based first on carbonic acid, and secondly upon humidity, lead to similar conclusions in each case.

There are, however, other sources of impurity in air. Dust is always present ; particles of inorganic dust, whether saline or carbonaceous, afford nuclei to which aqueous vapour has more or less affinity, and which afford the basis of haze, fog, and rain. The organic dust contains spores, moulds and bacteria, which are present everywhere, and of which some appear to be concomitants of disease, whilst others are the universal scavengers who convert the dead organic matter into the condition necessary to fit it to become again food for various forms of organic life.

It is difficult to base any calculation as to the purity of air in a room upon these minute organisms, because the condition of the air of a room in respect of its movement or quiescence largely determines the numbers which would be found at a given time in a particular part of the room.

Consequently the ventilation of a room for practical purposes must be based upon those conditions of the air which are capable of more accurate measurement, viz. the presence of CO_2 and of moisture.

In a warm climate the natural changes of temperature, and consequent alteration of the conditions of the movement of air, differ widely from those in temperate and cold climates.

In our temperate climate, a careful practical examination of the condition of barrack-rooms and hospitals, judged of by the test of smell, shows that arrangements which appear to provide for a volume of air much less in amount than that obtained by calculation will keep the room in a *fair* condition.

These results have pointed to about 1200 cubic feet of air admitted per hour in barrack-rooms occupied by persons in health. This need not be set down to errors in calculation or in theory.

There are many data which cannot be brought into the theoretical calculation.

For instance, the carbonic acid disappears in a newly-plastered or lime-washed room, and could be recovered from the lime, therefore a newly cleaned lime-whited room will present different conditions from a long occupied dirty room. Quicklime washing destroys fungi in dirty walls, as also does sulphurous acid fumigation. Now air has the same property, especially dry air ; and hence opening windows, turning down beds, and all such measures, act directly on the subsequent state of the air. Therefore an enormous effect is produced on all the elements of the above calculation if the windows of a room are kept open for several hours a day, instead of being closed.

Besides this, the conditions under which the air flows in and out of a room are so varied. The walls and ceiling themselves allow of a considerable passage of air. Examples of the porosity of materials have been already given. The ceiling affords a ready instance of porosity; an old ceiling is blackened where the plaster has nothing over it to check the passage of air, whilst under the joists where the air has not passed so freely, it is less black. On breaking the plaster, it will be found that its blackness has arisen from its having acted like a filter, and retained the smoky particles while the air passed through.

Moreover, the porosity of the walls materially influences the moisture, for a porous wall may absorb much moisture; and on this account rooms with walls of polished impervious material require much more air to pass through them. In the absence of sufficient ventilation, when the walls are colder than the air, moisture condenses on the walls.

Ill-fitting doors and windows allow of the passage of a considerable quantity of air.

In a temperate climate, where the changes of temperature of the outer air are rapid and considerable, these means of producing the outflow of air from and the inflow of air into a confined space are in constant operation. A sleeping-room is very warm when occupied at night; a rapid fall of temperature occurs outside, and at once a considerable movement of air takes place.

The majority of occupiers of sleeping-rooms in England close their windows at night; they also often block up the chimney by a register or otherwise, to prevent the blacks falling. These rooms have no special inlet or outlet for changing the air. In the morning they would no doubt come under Dr. de Chaumont's definition of 'very close'; and if it were not for the continual insensible change of air which passes through the walls, and the door and window chinks, &c., the occupants would be asphyxiated. A well-built house, unprovided with special means for the inflow of fresh air, is from the very completeness of its construction a real source of danger.

For these reasons, the form of a building is important, especially where rooms have to be occupied by large numbers of persons.

The air, thus insensibly coming in, should be taken from pure sources. Thus, barrack-rooms with outside walls are better than rooms opening out of a corridor, or on each side of a corridor. The air in a corridor becomes, after a time, saturated with impurities, and the interchange of air from it to the barrack-room becomes in time only an interchange of impure air. This is especially noteworthy in hospitals, where fresh air is of even more importance. The following, Fig. 4,

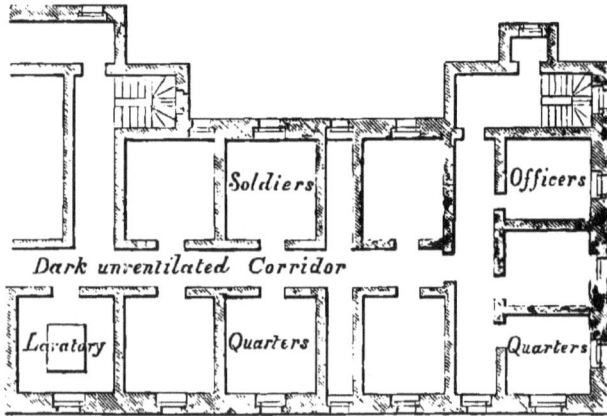

Fig. 4.

shows an arrangement of barrack-rooms built between the years 1850 and 1860, which illustrates this point; the outer wall affords only a quarter of the whole wall-space.

Fig. 5 shows the form of rooms which have been adopted in all recent barracks, in accordance with the principles here laid down. In this case the windows afford means for sweeping the bad air out of the room, so that the occupants shall have the opportunity of starting every day with fresh air; while the length of wall exposed to outer air, as compared with the inside walls, is as four to one.

These considerations bear essentially upon the construction of buildings occupied by large numbers of persons, such as barracks, workhouses, schools, or asylums ; but they are especially applicable to hospitals, where, as has been already shown, the emanations from a given number of sick are more perceptible than those from individuals in health ; and for this reason, in addition to numerous other reasons, the pavilion form of hospital construction presents advantages over other forms. In private dwellings the same conditions of occupation do not prevail ; and in these, therefore, considerations of comfort, and convenience of internal arrangement, may be allowed to have more weight in the design than the outer form of the building.

The foregoing observations will have shown that whatever be the cubic space, the air may be assumed to attain a permanent

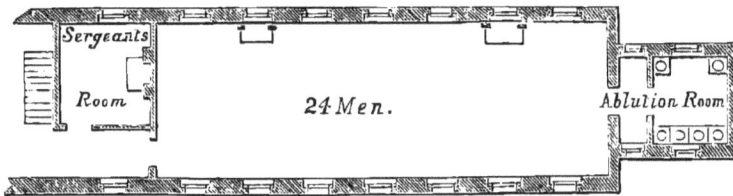

Fig. 5.

degree of purity, or rather impurity, theoretically dependent upon the rate at which emanations are produced, and the rate at which fresh air is admitted ; and that therefore the same supply of air will equally well ventilate any space, but the larger the cubic space, the longer it will be before the air in it attains its permanent condition of impurity. Moreover, the larger the cubic space, the more easily will the supply of fresh air be brought in without altering the temperature, and without causing injurious draughts.

One of the chief difficulties of ventilation arises from the draughts occasioned thereby. Every one approves of ventilation in theory ; practically no one likes to perceive any movement of air.

Large rooms, in addition to the advantage afforded of enabling the air to be changed with more comfort to the occupants than small rooms, also present the advantage of a larger wall-surface, and of more numerous windows, which allow of a larger insensible ventilation ; thus larger rooms will have, even in proportion to their occupants, an apparently less degree of impurity than small rooms. Although the uniform diffusion of carbonic acid is comparatively rapid in the air of a room, the organic emanations given out do not in practice diffuse themselves either rapidly or uniformly. They hang about in corners where there are obstructions to the flow of air, or near the ceiling, in which case they cool and fall down, and mix with the air of the room, thus increasing the impurities in the lower part of the room. Consequently there is no advantage in mere height in a room unless combined with means for removing heated air from the upper part. Indeed a lofty room with a space above the top of the windows or ventilating openings to which air loaded with emanation can ascend, remain stagnant, cool, and then fall down, is a positive disadvantage.

In a room with more than one occupant, it is necessary that a certain floor-space should be allotted to each occupant, for the purpose of allowing the currents of air to remove the emanations from one occupant without interfering with his neighbour, and to prevent the inconvenience of too close juxtaposition. The cubic space for soldiers in barrack-rooms occupied by day and night was fixed by the Royal Commission of 1858 at 600 cubic feet per man ; if we assume a room to be 12 feet high, that allows 40 superficial feet per occupant, and if the room be 20 feet wide, with beds on each side, the width across each bed, that is the linear bed-space, would be 5 feet.

In a room 10 feet high and 20 feet wide, a width of 6 feet of linear bed-space would be afforded ; but with rooms higher than 12 feet, it would be unadvisable to diminish the floor-space ; thus the floor-space necessarily to some extent governs cubic space. In warm climates a larger cubic space is given, mainly with the object of obtaining a larger floor-space. In hot climates as much as 100 square feet, and in some cases in temperate climates 80 feet of floor-space per occupant has been given in barracks, dependent on local conditions of the healthiness of the site, and of the plan of the buildings.

The Royal Commission on Cubic Space in Workhouses considered that where so many persons have to be lodged at the expense of the ratepayers, it was necessary to exercise the most rigid economy of space, and to supplement the deficiency of space by the strictest attention to ventilation and warming ; and they reported, that for dormitories in workhouses occupied only at night by occupants in health, 300 cubic feet would be sufficient, provided the wards did not contain more than two rows of beds, and that the height, if above 12 feet, was not reckoned in the calculation. This would allow a minimum floor-space of 25 feet per occupant, or with dormitories 17 feet wide, a bed-space of about 3 feet. This allowance is based upon the assumption that the domitories are only occupied at night, and that the most watchful care is bestowed on the efficiency of the ventilation.

In hospitals the cubic space must practically be dependent on the floor-space, for on this depends the distance of the sick from each other, the facility of moving about the sick, shifting beds, cleanliness, and other points of nursing. If there be a medical school attached to the hospital, the question of area has to be considered with reference to affording the largest amount of accommodation practicable for the teacher and his pupils.

A ward with windows improperly placed, so as not to give sufficient light, or where the beds are so placed that the nurse must necessarily obstruct the light in attending to her patients, will require a large floor-space, because the bed-space must be so arranged, and of such dimensions, as to allow of sufficient light falling on the beds. In well-constructed wards with opposite windows, the greatest economy of surface area can be effected, because the area can be best allotted with reference both to light and to room for work.

In a ward 24 feet in width, with a window for every two beds, a 7 feet 6 inch bed-space along the walls would probably be sufficient for nursing purposes. This would give 90 square feet per bed. And as a rule, if the locality is healthy, the floor-space may be fixed at about 90 square feet per bed in wards of a general hospital where average cases of sickness are treated in this climate, with the understanding that the area shall be increased for surgical cases, or if the building is designed for a medical school, or where from unavoidable circumstances an unfavourable site must be selected.

The practice in regard to area differs considerably in different hospitals. In the naval hospitals about 78 square feet per bed is allowed. In the Herbert Hospital, where there is no medical school, 99 square feet per bed. The cubic space which results from this latter area, with wards 14 feet high, is 1260 cubic feet.

In the Royal Victoria Hospital at Netley, where there is a medical school, it is 103 square feet. In St. George's Hospital it is about 70 square feet.

From this minimum, it varies to 138 square feet in Guy's Hospital. In the new Hôtel Dieu at Paris, the space per bed is from 104 to 110 square feet, and in the new St. Thomas's Hospital it is 112 square feet. This latter area is considered sufficient both for nursing and teaching purposes.

In the new fever hospitals, and in wards for bad surgical cases, where the emanations from the patients are considerable, it is found desirable to afford a larger floor-space, varying from 150 to 200 superficial feet per bed, and to afford a lineal bed-space of 12 feet; in cases of diphtheria 15 feet. This entails an enlarged cubic space.

In cases of fever, if a separate ward is not available, a bed should be removed on each side of the fever patient.

In workhouse hospitals, where the strictest economy is sought, and where the cases are generally of a more chronic character than in ordinary hospitals, the Royal Commission on Cubic Space in Workhouses required 850 cubic feet per inmate, with a minimum of 70 square feet of floor-space, and a clear space of six feet across each bed, and that no bed should be placed in the middle of the floor.

In special workhouse hospitals for fever and small-pox patients, 2000 cubic feet, and a minimum floor-space of 156 square feet, is provided.

For lying-in wards in workhouses a minimum of 1200 cubic feet and 100 square feet is the standard : but these wards in workhouses are seldom continuously occupied.

In wards partially occupied by day and by night for aged, chronic, and infirm cases, with the use of a day-room, 500 cubic feet and 42 superficial feet were specified as necessary ; on the condition that the ventilation would be adequate and carefully watched.

Dormitories in schools should not afford less floor-space than from 50 to 60 square feet per occupant. In school-rooms a floor-space of from 25 to 35 superficial feet has been considered sufficient, on the understanding that the room is occasionally emptied and the air renewed : but with the large number of pupils in elementary schools it is difficult to maintain adequate purity of air without some form of mechanical extraction or inflow of air.

In prison cells, where the prisoner is confined continuously, the superficial area per occupant should not be less than from 80 to 120 feet, according to the character of the prison and other circumstances.

Thus the proportion of floor-space and cubic space in any room must be regulated to a certain extent with reference to its shape and to the conditions of its occupation, as well as to its capacity for ventilation.

CHAPTER VII.

AIR of the composition before mentioned, viz. 210 of oxygen to 790 of nitrogen, is a heavy body. At a temperature of 32°, and with the barometer at 30 inches, which is about the mean sea level, dry air weighs 566·85 grains per cubic foot. The pressure of the atmosphere on any surface is nearly 14·7 lb. to the square inch ; and a column of air of about 87·6 feet in height, under these conditions, will balance a column of mercury ·1 (or one tenth) of an inch in height.

The molecules of air are but feebly attracted to each other, and small increases of temperature, or slight diminutions of pressure separate the particles from one another, and thus one cubic foot of expanded air weighs less. Similarly, small decreases of temperature bring the particles nearer together, and make the cubic foot of cold air heavier than the standard above mentioned. This expansion and contraction are equal for equal increments or decrements of temperature.

This increase of volume amounts to 0·365, or about three-eighths of the original bulk, in the process of being heated from the freezing to the boiling point of water ; or nearly ·00203 for every degree of Fahrenheit.

Thus, if the air inside a room were 20° Fahrenheit warmer than the air outside, the air in the room would be expanded to a 25th part more in bulk, and would to that extent be specifically lighter than the outside air.

This dilatation of air by heat and its contraction by cold are expressed by the formula $M_1 = (1 + a\,t)\,M$

when M = volume at 32° and the barometer at 30 inches,

M_1 = volume at the temperature of t degrees above 32°,

a = co-efficient derived from experiments on the proportion of the increase of volume of air for each degree of elevation of temperature ($= \cdot 00203$ for each degree of Fahrenheit).

When temperature is decreasing the formula is

$$M_1 = (1 - a\,t)\,M.$$

When the temperature of air and the space it occupies increases, its density—that is, its weight per cubic foot—decreases in the ratio expressed in the following formula, assuming barometric pressure constant,

$$d = \frac{d_1}{1 + a\,t}.$$

The following table shows the density of air at different temperatures.[1]

Weight of air per cubic foot under 30 inches pressure of Mercury.

Temperature Fahrenheit.	Dry Air.	Air saturated with vapour.
	Grains.	Grains.
0°	606·37	606·03
20°	581·05	580·26
32°	566·85	565·58
40°	557·77	556·03
50°	546·82	544·36
60°	536·28	532·84
80°	516·39	509·97
100°	497·93	486·65

It follows that since warmed air expands and becomes

[1] The weight of a cubic foot of air at different temperatures and pressures may be found approximately by the formula weight in lbs. =

$$\frac{1.3253 \times \text{Height of Barometer in inches of mercury}}{459 + \text{Temperature Fahrenheit}}$$

lighter, and since cooled air contracts and becomes heavier, the colder air has a tendency to press the warm and lighter air upwards, and to occupy the place of the warmer air. In an enclosed space, the rate at which the warmer air will thus be pressed upwards by the inrush of colder air, and be forced out, will depend upon the form, size, and materials of the openings which permit its escape.

Everywhere the heating and cooling of the air is going on; the sun's rays, the proximity of a warm body, the vicinity of a cool shaded surface, all cause changes of temperature, and thus create movements or currents in the air. It is on this law of the dilatation of air that all the movement of air depends, from the winds and hurricanes to the ventilation of houses, except where air is propelled by fans or by other mechanical appliances.

The change of air in a confined space may be the result of the movement of the outside air—or of difference of temperature between the air inside and that outside.

The movement of outside air across a building will tend to draw air out of chimneys or air-flues; or out of openings to leeward of the building.

In a room, as air is warmed by the bodies of the occupants, it ascends; and would pass away through air-flues or chimneys; the air which comes in contact with the colder walls of the room or with the glass of the windows, cools, and falls down. A draught experienced near the window, does not necessarily show that air is coming in through the window, it may simply result from the cooled air which is falling.

It is also noteworthy that air saturated with vapour is lighter than dry air, and air will therefore flow upwards more easily in proportion as the ascending column is saturated; and therefore breathed air which contains moisture from the occupants of a room ascends more easily than the drier unbreathed air.

The law which regulates the movements of the air in a confined space, when the temperature is higher than that of the outside air, depends upon the following considerations :—

1. Upon the difference of temperature of the air inside the confined space, as compared with that outside.

2. Upon the area and other conditions of the apertures through which the warmed air can flow out and the cooler outside air can flow in to take its place.

3. Upon the height of the column of ascending warmed air.

If AB represent the height of a column of air of the outside temperature t_1, and AC the height of a column of the same quantity of air expanded by the warmer temperature t, then the velocity at which the warmer air ascends will be that which would be acquired by a body falling from C to B.

That is, $V = \sqrt{2g \times BC}$,

if V=velocity of ascending air in feet per second ;

H=height of shaft in feet ;

t=temperature in shaft ;

t_1=temperature out of doors ;

a=the coefficient of dilatation of air, which for $1°$ Fahrenheit=·002036 ; and for $1°$ Centigrade=·003665 ;

g=32·17.

The theoretical equation becomes $V = \sqrt{2gHa(t - t_1)}$.

Therefore the velocity, and consequently the volume of air varies with the square root of the difference between the temperature inside and outside the shaft.

The actual movement of air in a chimney is very different, owing to the resistance from friction to which the moving air is subjected. The friction varies directly with the square of the velocity of the air-currents, and with the length of the channel or flue, and inversely with the diameter or area of the flue ; and is, moreover, much influenced by the material

F

of which the sides of the flue are constructed. With a sooty
flue, or a flue with rough sides, the velocity, with equal temperatures, has been found to be one-half that of a smooth
clean flue.

The velocity is, moreover, diminished by the friction caused
by impediments to the ingress of the fresh air required to
supply the place of that which flows out ; and therefore an
efficient system of ventilation requires that the extraction of
air should be accompanied by convenient arrangements for
the supply of fresh air to take its place, and vice versâ.

If these resistances be represented by the constant K, to be
determined for each case depending on the form and material
of the shafts and air-channels, and on other considerations,
such as their freedom from dirt, &c., the formula may be
generally expressed as follows—

$$V = K \sqrt{(t - t_1)H\, 2ga}.$$

Péclet, in his treatise on the application of heat, has given
the following formula to include some of the resistances—

$$V^2 = \frac{2ga\, H(t - t_1)D}{D + 2gHK};$$

where D = diameter of shaft, circular flue, or square root of
area of rectangular flue ; K = the coefficient of resistance.

And he determined the coefficient of resistance, corresponding to this formula, due to pottery chimneys to be ·0127 ;
sheet-iron chimneys to be ·005 ; and cast-iron chimneys to
be ·0025. This formula gives rather too high results. Phipson
suggested

$$V^2 = \frac{DH(t - t_1)}{L + 16D}$$

where L = length of evacuation channels. Hurst gives

$$V^2 = \frac{·13\, DH(t - t_1)}{D + KL},$$

where the dimensions are in feet, and K = ·02 for clean glazed
earthenware flues, ·03 wood flues, ·06 sooty flues. The difficulty of obtaining a uniform coefficient for the resistances will

be made apparent from the fact that in flues of the size of ordinary chimneys, soot or accumulations of dust on the sides seriously affect the velocity; General Morin found that the presence of a cobweb in a flue almost entirely checked the passage of air. The main conditions to be attended to in the design of air channels are that they should be as straight as possible, with smooth sides, and with such an area as will prevent the necessity of maintaining a high velocity in the channel.

The circumstances which affect the flow of air are thus so varied that it is preferable, in estimating the amount of air removed for purposes of ventilation from buildings already constructed, to measure the actual volume of the air in the flues or air-passages ; that is to say, to cause it to pass along a channel—the size and area of which are known—and then to measure the velocity with which the air passes through this channel. The multiple of the area into the velocity in a given time gives the volume which passes through in that time.

There are various ways of measuring the velocity. It may be measured by puffs of vapour of turpentine; by balloons filled with hydrogen, and weighted to be of the exact specific gravity of air. In ordinary cases, the most convenient method is by means of an anemometer. An ordinary form of anemometer is that of vanes fixed to a spindle, the revolutions of which are recorded by a counter. The vanes are turned by the direct action of the current of air, and the number of revolutions which are recorded by the counter gives the velocity. Of course the value of these revolutions has to be ascertained in the first place by direct experiment ; that is, by forcing a known bulk of air, at a uniform rate, through a channel of a given size ; and ascertaining the number of revolutions made by the vanes. The most convenient apparatus for this purpose is a graduated vessel constructed on the principle of the ordinary gas-holder, arranged to move with a uniform speed, from which a known quantity of air can be expelled at will through a channel of convenient dimensions in connection with it.

For anemometers of this pattern the formula which must be applied to ascertain the velocity takes the following form—

$$V = a + bN,$$

where $V =$ velocity of air ;

$a =$ a constant number showing the minimum speed of current which will move the vanes, and which should be contrived to be as small as possible by means of light vanes and delicate machinery ;

$b =$ a constant coefficient ;

$N =$ number of turns of the spindle in 1 second.

Fletcher's Anemometer is another very convenient form for measuring the speed of air in heated flues.

The instrument consists of two parts ; firstly, of two metal tubes of about $\frac{3}{10}$ of an inch internal diameter, open throughout, and of any length ; secondly, of a manometer, or pressure-gauge. Of these tubes, the end of one is straight and plain, while that of the other is bent to a right angle. When in use these tubes are placed parallel to each other, and so that their ends are exposed to the current of air to be measured. They lie at right angles to the current, which thus crosses the open end of the one, and blows into the bent end of the other.

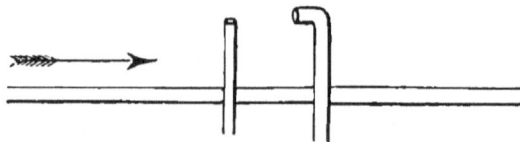

Fig. 6.

By this means a partial vacuum is established in the straight tube, whilst the pressure of the current forces the air into the bent tube ; a differential manometer, attached to the outer ends of the tubes, shows the excess of pressure in the bent one over that in the straight one.   The manometer

used is a simple U-tube set vertically, containing ether, fitted
with microscopes and vernier scales, by which the difference
of level of the surfaces of the ether in the two limbs can
be measured to $\frac{1}{1000}$ of an inch. This difference of level
between the columns of ether becomes a measure of the
speed of the current passing the ends of the anemometer
tubes; the law which governs the speed is expressed generally
by the formula $v = \sqrt{p} \times 28.55$. The corrections to be made
for small variations of barometric pressure and temperature
are unimportant. The corrections when required are embodied in the following formula—

$$v = \sqrt{p \frac{h}{29.92} \cdot \frac{519}{459+t}} \times 28.55,$$

where p is the height of the column of liquid driven up the
tube measured in inches, and v is the velocity measured in feet
per second of air at a temperature of t degrees Fahr., under
a pressure of h inches of mercury.

Tables of the velocities corresponding with the readings are
supplied with the anemometer, and also a table of correction
for temperature.

The variations of temperature to which the manometer
itself is exposed are not great, being those of the external
atmosphere only.

This is a very convenient form of anemometer, because the
pressure-tubes may be of any length or diameter, and may
be connected with the manometer by india-rubber tubing of
any length. Therefore the readings may be observed in any
convenient place at a distance from the flues.

The pressure-tubes should project into the air-current to
the extent of one-sixth part of its diameter, to measure the
average velocity. It is desirable, after observing the height of
the ether in the tube, then to reverse the connection with the
manometer, and to take a second reading; if the smaller
reading be deducted from the greater, twice the height of

the column supported by the difference of pressure is obtained, and the error of observation is halved. But in this, as in the anemometer with vanes, there is a difficulty in accurately observing low velocities, i. e. under 1 foot per second.

It is worth noting that a sheet of light tracing-paper moved through the air at 2 feet per second takes up an angle of 45°, and affords a ready means of measuring that velocity; and for smaller velocities the angle assumed by the flame of a candle affords a fairly accurate index according to the following table.

Velocity of flow of air. Feet per second.	Angle of inclination of flame of candle with horizon.
1.6	30°
1.0	40
0.75	50°
0.50	60°
.40	65°

In a shaft open at the top, containing air at a temperature above that of the outside air, and in which means exist for keeping the incoming air at a similar uniform temperature above that of the air outside, the velocity of the upward current will vary with the height of the shaft. In order, therefore, to effect the extraction of the air in an enclosed space with the least expenditure of heat, the shaft should be made as high as possible. Any diminution in height or in area must be compensated for by additional heat in the shaft ; this means an additional consumption of fuel to keep up the temperature.

Therefore the economical application of the law of dilatation of air depends on the height available for the shaft, upon its area, and upon its being as free from friction and other resistances as possible, and upon the difference of temperature which can be obtained indoors and out. It will thus to some

extent depend upon climate : where the outer air is cold, and
a comparatively high indoor temperature must be maintained
for comfort, it is most economical ; but where the difference
between the indoor and outside temperature is small it is
less economical. Thus General Morin found that whilst in
winter the ventilation in the *Conservatoire des arts et métiers*
by extraction-shafts was continuously maintained with an
expenditure of 1 lb. of fuel for 8700 cubic feet of air removed,
in summer 1 lb. of fuel removed only 3000 cubic feet.

Arrangements for the change of air in a confined space
which depend on difference of temperature are thus the
simplest form of ventilation. Such arrangements can, of
course, be applied only where there is a difference between
the inside temperature and that outside—whether the air
inside is warmer or cooler than that outside. But where
fuel must be used solely for obtaining movement of air, it
can be applied more economically in the propulsion of air
by mechanical means.

Propulsion of the air by means of fans or of pumps may
be used either to force the air into a room, or to extract
it from a room.

Theoretically, the propulsion of air into a room would expel
all the foul air through the cracks of windows and doors, even
if no special apertures were made for its removal, and the
existence of pressure in the room would tend to prevent
draughts of cold air from doors and windows ; but in practice,
in the ventilation of hospital wards, the system of propulsion,
i. e. forcing the fresh air into the room, and allowing the
vitiated air to find its way out, has not been generally found
successful as a means of purifying the air. The air forced in
seems to seek the first place of escape, and unless the system
is combined with an efficient system of extraction, much of
the vitiated air will remain in corners and dead angles. It is
therefore advisable always to combine with a system of propulsion for the inflowing air some method of extraction of

vitiated air. Where the circumstances allow of it, it will be found simpler to dispense with propulsion, and to rely upon the action of extraction-shafts to draw in the air required through adequate channels provided for the ingress of fresh air. In cases, however, where a large volume of air is required to be passed continuously into a confined space, and the channels are limited in size, it may be found advisable to assist the movement of the inflowing air by propulsion.

The compression of the air caused by propulsion raises the temperature of the inflowing air. Some experiments of General Morin showed that in a system of ventilation where the pumping in of the air gave a pressure equivalent to 2 inches of water, the temperature was raised from 20° to 25° Fahrenheit, between the temperature at the place from which the air was drawn into the fan and the temperature at the inlet into the room.

The system of extraction by fans is of the highest value in cases where it is desired to remove particular impurities with great rapidity ; such, for instance, as in crowded schoolrooms or in workshops where the dust of cotton, steel filings, or injurious emanations produced locally are sought to be removed at once.

The friction of air varies as the square of the velocity multiplied by the pressure against the sides of the passage. This pressure being uniform, its total amount depends upon the total surface ; that is, the length multiplied by the perimeter of the cross section. The force required to propel air through any passage is therefore equal to the square of the velocity into the total surface multiplied by the coefficient of friction. It is more convenient to state the force in lbs. per square inch, or per square foot, or as so many inches of water pressure.

The best form of the formula for practical purposes of ventilation seems to be—

$$H = \frac{K V^2 P L}{A},$$

where

$H=$ head of pressure in feet of air of same density as the flowing air ;

$L=$ length of the pipe or passage in feet ;

$P=$ perimeter of cross section in feet ;

$A=$ area of pipe or passage in square feet ;

$V=$ velocity taken in thousands of feet per minute ;

$K=$ coefficient of friction $=0.03$; but this varies according to the nature of the sides of the passage.

This formula is perfectly general, and may be used for any fluid ; H always being the head, stated in feet, of the flowing fluid.

The weight of 1 foot of air, under the pressure of one atmosphere, at $32°$ Fahrenheit, is equivalent to 0.0807 lb. per square foot. The weight of 1 inch of water at $39°.1$ Fahrenheit equals 5.2 lb. per square foot ; therefore 1 foot of air, at $32°$ Fahrenheit, under the pressure of one atmosphere, exerts a pressure equivalent to 0.0154 inches of water.

For circular passages, taking D for the diameter, the formula becomes—

$$H=K\,V^2 \times \frac{4L}{D}.$$

These formulae are only applicable to passages whose diameter is small in proportion to their length. For short passages the actual length should be increased, for purposes of calculation, by about 50 diameters of the passage : thus the formula for short circular passages becomes

$$H=K\,V^2 \times \frac{4(L+50D)}{D};$$

for short irregular-shaped passages,

$$H=K\,V^2 \times \frac{PL+200A}{A}.$$

It is beyond the limits of this treatise to enter into the question of the form of propeller which should be used. It may however be observed, that the best fans have produced from 70 to 75 per cent. of useful effect in proportion to the power employed.

There is a third system, which can only be applied in special cases—viz. that of ventilation by means of compressed air, adopted in the Mont Cenis and St. Gothard tunnels during construction. The compressed air, after having been utilised to work the boring machines, escaped into the tunnel, and provided fresh air for ventilation.

Compressed air has also been applied to produce a current, and thus to extract vitiated air, somewhat on the principle of the steam jet which causes the draught in a locomotive chimney, but acting by its momentum only.

If M = the volume;

V = the velocity of the air injected into the channel;

M^1 = the volume of the additional air drawn into the channel by the movement of the injected air;

$M + M^1 = M_1$ = the total volume of air passing along the channel;

V_1 = mean velocity of the air, as it flows along the channel;

$$MV \quad (M + M^1) V_1 = M_1 V_1.$$

If d = density of injected air;

D = density of air in channel;

s = section of orifice of injector;

S = section of orifice of channel;

$$M = \frac{ds V}{g} \quad \text{and} \quad M_1 = \frac{DS V_1}{g}.$$

The system was used for the ventilation of some of the galleries of the Exhibition at Paris in 1867, and was subjected to experiment by General Morin; and he shows that the useful effect varies inversely with the cube of the velocity of the air from the injector.

In a channel having a diameter of about 8 feet, with a jet of injection of about 5 inches diameter, the velocity in the channel being from 6 to 9 feet per second, and the velocity of

the jet of injection about 220 feet per second, the useful effect
of the jet only equalled $\frac{1}{25}$ of the power of the jet ; and as
the jet only utilised half the useful effect of the engine, the
general result was to utilise only $\frac{1}{36}$ of the motive power
expended. Therefore this method of propelling air is not
advantageous; but it may be resorted to in special cases
where difficulties exist in applying other methods of propulsion.

There remains the method of extraction by the movement
of the atmosphere alone. If an open tube or shaft be carried
up from a room or enclosed space to a point above, where
the top is exposed to the free movement of the atmosphere,
an upward current will prevail in the shaft so long as there
is a movement in the atmosphere. The movement is of
course unequal in its action. It is powerful when the wind
is high. In calm weather it is very small; but in this
country, as already mentioned, the average velocity of the
atmosphere is above 17 feet per second, and it is rarely
quite at rest.

It is very difficult to measure the relation which the current
in a tube or shaft caused by this method of extraction bears
to the velocity of the wind, because there are so many conflict-
ing elements to be considered. The formulae for calculating
the velocity of wind in some of the standard anemometers
are not entirely satisfactory for the very low velocities, because
the constant factor which increases as the velocities diminish
has not yet been satisfactorily determined. Also the action
of wind, whilst it tends to exhaust the air through the tube.
is, at the same time, acting on all other openings in the
building, either to exhaust or to force in air. Hence gusts
of wind will sometimes cause a reverse action in the tube,
in consequence of some other opening acting temporarily as
a means of extraction.

The temperature inside and outside must also be considered.

If the atmosphere should be without perceptible movement

in cold weather, when the temperature indoors is maintained for comfort above that out of doors, the difference of temperature will cause an upward movement in the shaft. In hot weather, if in still weather the shaft is colder than the outer air, a down current may ensue; but if, in hot weather, there should be little or no movement in the shaft, this occurs at a time when the windows can be kept open, and the air be renewed by this means.

The friction in the shaft varies inversely with the area; and with small tubes it forms a very perceptible element of retardation. Experiments made with tubes three inches in diameter tend to show that the velocity obtained in the tube was about $\frac{2}{3}$ of that of the wind; larger diameters, on the other hand, produce velocities of from $\frac{1}{2}$ to $\frac{2}{3}$ the velocity of the wind.

These results were obtained in a place supplied, as far as possible, with fresh air to replace that removed, in a manner independent of the movement of the atmosphere, and with the top of the tube freely exposed on all sides.

In consequence of the numerous causes of disturbance enumerated above, this method of extraction, although very powerful in windy weather, could not be relied on to act on all occasions with certainty as an extraction-shaft. But it can be relied on to ensure in one way or other, and to a certain extent, a continual change of air.

A tube or shaft with an open top acts best. It is, however, necessary to protect the top, to prevent rain from entering the tube; if this cover or cowl afford an area of outlet not greater than that of the shaft it will tend more or less, according to its shape, to delay the current in the tube or shaft. But if the cowl afford a larger area of outlet than the shaft it may accelerate the current. The efficiency of so-called air-pump or exhaust cowls depends upon the proportion which their area and connections bear to that of the flue or shaft. A cowl with a curved head, gradually enlarged at its mouth into an oval shape, with an aperture somewhat larger than the flue or tube, arranged to move round with the wind. so as always to present its back to the wind, would appear to be the form of covered top best adapted to facilitating extraction. Of the fixed cowls Boyle's is very efficient.

CHAPTER VIII.

PRELIMINARY CONSIDERATIONS UPON THE INFLOW AND REMOVAL OF AIR.

THE efficient ventilation of a building depends upon the efficient ventilation of each room in that building. The first matter for consideration therefore is the manner in which the ventilation of a room can be best secured.

The most efficient way of thoroughly renewing the air of a room is by means of open windows and doors, when these are so placed as to ensure a thorough draught ; and open windows should always be resorted to when the weather permits.

But the windows cannot always be kept open. In cold climates the windows must be closed to keep the rooms warm. In hot climates the windows must often be kept closed to keep the rooms cool. In the one case warmed air, in the other cooled air, should be admitted independently of the windows. Moreover, windows are so placed in a room as to meet the requirements of light, and do not therefore necessarily occupy the most advantageous position for the continuous admission of air. Therefore every room should have special arrangements for the admission and extraction of air. This entails inlets and outlets, and in some cases ducts and flues leading thereto.

The rate and temperature at which air is removed and admitted materially affects the comfort of the occupants of a room.

The velocity of the air as it flows in and out of a room, as measured at the openings for admission or exit, should not exceed 1 foot, or at most 2 feet, per second ; firstly, in order to prevent a sensible draught being felt ; and secondly, because a low velocity is favourable to the uniform diffusion of the incoming air through the room.

To avoid friction, it is convenient that the velocity in the channels leading to the main extracting shafts should not exceed from 3 feet to 4½ feet per second, and the velocity in the main extracting shafts themselves should not exceed from 6 to 7 feet per second.

This latter velocity will, under general circumstances, and where the extraction is effected by means of a heated shaft, be provided by a difference of temperature between the inside and outside of from 30° to 35° Fahr. In special cases, on grounds of construction or otherwise, it may be found necessary to exceed this.

These velocities would be regulated by the size given to the inlets, outlets, supply channels, and extracting shafts, as compared with each other respectively, and with the quantity of air to be supplied and removed ; which quantity would depend upon the number of occupants of the rooms to be provided for, and the amount of air to be allotted to each ; on the conditions required for the ventilation of corridors, lobbies, staircases, water-closets, and other subsidiary accommodation ; and on any other conditions necessary to be adopted in fixing the supply. The areas thus obtained should be the free areas, exclusive of gratings or other impediments.

Air should be introduced and removed at those parts of the room where it would not cause a sensible draught. Air flowing against the body, at or even somewhat above the temperature of the air of a room, will cause an inconvenient draught ; from the fact that as it removes the moisture of the body, it causes evaporation or a sensation of cold.

Air should not as a rule be introduced near the floor level. The openings would be liable to be fouled with sweepings and dirt.   The air, unless very much above the temperature of the air of the room, would produce a sensation of cold to the feet. The orifices at which air is admitted should preferably be above the level of the heads of persons occupying the room, an advantageous position is at from 7 feet 6 inches to 9 feet above floor level; the current of inflowing air should be directed towards the ceiling, and should be as much sub-divided as possible by means of numerous orifices.

Air admitted near the ceiling very soon ceases to exist as a distinct current; and will be found at a very short distance from the inlet to have mingled with the general mass of the air, and to have attained the temperature of the room, partly owing to the mass of warmer air in the room with which the inflowing current mingles, partly to the action of gravity; where the inflowing air is colder than the air in the room, and where there are open fireplaces this mingling is acce-lerated by the action of the fire.

The general form of a house affects the ventilation of the rooms.   In a house with a central staircase carried from the ground floor to a lantern light in the roof, when the air in it is warmer than the outside air, it becomes a powerful shaft; and being of such a large size, tends to draw in the air from adjacent rooms, and even down the chimneys, especially when the chimney is cold, or when its size is greater than required for the removal of the air provided for its supply.   In the latter case, a double current, one up, one down, and both sluggish, is sometimes established in the chimney.

A smoky chimney will therefore often be cured by being reduced in area, so as to reduce the volume of air required to fill it, or by an adequate provision of fresh air to the room in which it is placed, or by a supply of air in an adjacent hall or staircase.

The effect of a high wind is to diminish the air-pressure in a house when all windows are closed on the side subject to the outer air-pressure: hence although some of the chimneys may have an accelerated draught due to the velocity of the wind, the diminished air-pressure which results in some cases may be to cause other chimneys to smoke, the remedy for which would be to open slightly the windows on the side of the air-pressure, so as to allow the pressure to act equally through the house. It will frequently happen that the open fires in the rooms which require to be fed with fresh air, and the warm air shaft of the central staircase, will combine to draw in the air from every available opening. They will draw up the air from the basement; they will draw it, if they can, from the water-closet and sink-pipes, unless the water-closets and sinks are in projecting buildings, cut off by lobbies ventilated from the open air.

For these reasons it is of importance not only to shut off the staircase from the basement, but to provide fresh air to the staircase; and when open fires are used, to supply each room in which there is an open fire-place with its own supply of fresh air. If the temperature of the room is to be kept up to a pleasant heat, this must to some extent be warmed air.

In a building with halls and passages, the inflow of cold air through the doors when opened must be guarded against by the independent supply of fresh air to, and the extraction of vitiated air from, all parts of the building, and by the maintenance of a proper temperature in all parts. By this means draughts occasioned from opening doors would be avoided. An inflow of cold air occurs only where the ventilation and warming are defectively arranged.

The fresh air should be obtained from places where there are no adjacent sources of impurity; especially not from the vicinity of ash-pits, manure-heaps, gully-gratings, or other sources of foul air. The inlets from the outer air should be at least two feet from the ground, and the surface near should be paved and sloped away from the inlet, so as to carry off wet rapidly. It has been estimated that the impurities of town air are very much diminished at 200 yards height, and are not found above 600 yards in height—a London fog will rarely be perceptible at a height of above 100 yards—consequently to bring pure air into a large town the simplest way would be to draw it down from a height by means of a high shaft or tower. The Houses of Parliament could be supplied with very much purer air than is now provided for them by bringing the air from the top of the Victoria Tower.

The temperature at which air should be delivered in a room depends somewhat on the temperature of the outer air. With a temperature out of doors of 86°, air delivered at 60° would be felt as too cold. Similarly, in the arctic expedition, with an out-of-doors temperature—40° below zero, a temperature of +40° was felt to be hot.

In this climate the temperature indoors should be maintained at from 58° to 66°, and the warmed inflowing air should be supplied at a temperature a few degrees above this, but as nearly as possible not to exceed from 68° to 75°, or at most 80°. The hygrometric condition of the air must also be considered. If the air is too dry, it may, after it has been warmed, be passed over vessels containing water, to allow it to take up additional moisture—but the natural dampness of an English climate renders this less necessary here than it may be elsewhere. It has not unfrequently happened that complaints of the oppressiveness of dry air artificially warmed have been caused rather by the insufficiency of the supply of warmed air than from its dry condition.

The channels for the admission of air should, as a rule, be short, direct, and accessible.

Long channels collect dirt, and form a refuge for insects. But if it is necessary to make them long, they should be easily accessible for cleaning. Deep underground channels and receptacles will modify materially the temperature of the inflowing air, because in winter the temperature of the earth within a short distance of the surface is warmer, and in summer it is cooler than that of the air ; channels for the purpose of utilising this effect of temperature should be of a rectangular shape, so as to afford a large surface in proportion to their area ; and they should be of a good conducting material.

Underground channels should always be impervious to ground air ; and, as has been already shown, there are very few materials impervious to air.

The liability of air in underground channels to mix with ground air would be diminished if the air were supplied to the channels or receptacles by propulsion, and retained in them under some pressure.

Underground channels should be perfectly dry. Damp channels overcharge the air with moisture; and thus they interfere with the warming of the air; they induce the presence of animal and vegetable life, which dies and decays, and renders the air impure.

In towns where the atmosphere is full of particles of soot and other impurities, the inflowing air deposits these particles

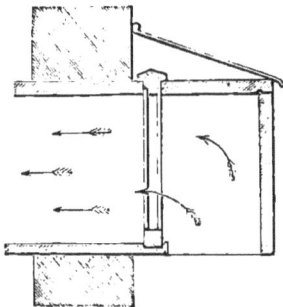

Fig. 7. Section of entrance to fresh air flue with cotton wool filter.

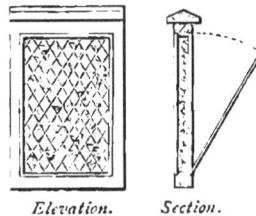

Fig. 8. Frame to hold cotton wool.

Elevation. *Section.*

and rapidly blackens any surfaces it impinges on. The impurities may be removed by passing the air through a filter made of cotton wool laid lightly, to a thickness of about half an inch, on a copper wire frame. (See Figs. 7 and 8.)

The cotton wool must be renewed at intervals, dependent on the state of the atmosphere. In London fogs it should not be left on for more than a few days. In clear weather in London it will last two or three weeks.

Sliced sponge acts equally as an air filter, and may be easily washed, dried and replaced.

A system has been introduced of endeavouring to remove impurities from the inflowing air by causing the current to come in contact with a surface of water. In damp weather this is liable to the inconvenience that the water keeps up the saturation of the air. In order to purify town air before its admission into hospital wards, Mr. Key has devised a screen of horsehair and fibre, through which the air is drawn by means of a fan. A sheet of water is poured automatically at short intervals down this screen. The wetted surface of the screen arrests impurities contained in the air; and these are washed off by means of the automatic periodic flush.

In dry hot weather the system which has recently been adopted for washing smoke by passing it through a chamber filled with spray, might be adapted for washing the impure air of towns.

Extraction-shafts, when for ventilation only, should be placed so as not to occasion unpleasant draughts; but their position must depend also upon other conditions, which will be further alluded to: meanwhile it may be observed that it is advisable that as far as possible extraction-shafts from different rooms, the operation of which depends on temperature, should be independent of each other. Where the rate of extraction is sluggish, and the temperature is low in the shaft, there may arise conditions in one room which may determine a reverse current, and then, when there is not a complete separation, the bad air from one room may be introduced into another room. And even when the flow of air into an extraction-shaft has been rapid through the outlets from all the rooms communicating with it, smells from one room have been diffused into other rooms along the line of the extraction-shaft.

Extraction-flues in which the temperature exceeds that of the outer air should be arranged whenever possible to conduct the air upwards, so as to assist its flow. Every descent in the flue has to be compensated for in some way, either by extra height of shaft, temperature, or expenditure of force.

Thus in a chimney-flue, where the extraction depends upon the heat in the flue if a portion of the flue near the fire is horizontal, it is found in practice that the flue cannot be depended on to act efficiently unless the vertical height is at least double the horizontal length.

The best form for an extraction-flue is the circular form, because it affords a maximum of area with a minimum of perimeter, or surface, and therefore causes a minimum of friction.

The various considerations enumerated above apply equally whatever be the manner in which the air is admitted or removed.

It should, however, be accepted as an axiom, that by the best ventilating arrangements we can only obtain in our rooms air of a certain standard of impurity ; and that ventilating arrangements are only makeshifts to assist in remedying the evils to which we are exposed from the necessity of obtaining an atmosphere in our houses different in temperature from that of the outer air. To obtain that temperature we sacrifice purity of air ; therefore, whenever the outer temperature permits it, windows should be widely opened, so as to replace the air of the room by fresh air, as often as possible.

CHAPTER IX.

SIMPLE VENTILATION WITHOUT WARMED AIR.

THE simplest way of obtaining a change of air in a room is to take advantage of the movement in the air produced by a change of temperature, or by the action of the winds.

Wherever there is an outlet and an inlet for air in a room, this system will operate. The open fireplace is one example of it ; the sun burner is another example, but the system is also applied in every room in which there is an opening at the upper part, out of which the warmed air can pass, and an opening either level with it or below it, through which fresh air can flow in.

Thus an ordinary sash window is the simplest example. If the top sash is lowered and the bottom sash raised, the warmed air passes out of the room at the top, and the cooler outer air flows in below. Hence, for the admission of air to an ordinary room, provided with a fireplace, but unprovided with special inlets, a very simple plan is to cut a slit above the lower bar of the upper sash of a window, so as to leave a clear space of about a quarter of an inch along its whole length, through which the fresh air will be drawn in in an upward direction.

Ventilators which act similarly to direct the current of air towards the ceiling are sometimes introduced into the upper panes of window-frames, such as hopper ventilators, or

Moore's louvred panes, but these are all makeshifts. Every room should have special inlets and outlets for air, arranged so as to be independent of the windows; although the windows, when they can be kept open, are the best means for the renewal of the air in a room. As an instance of this— assume a room with windows on opposite sides, of which the open part of each is 3 feet wide by 3 feet high, this affords an area of 9 superficial feet. With a movement of air across the building of only five feet per second, those openings would enable a volume of 160,000 cubic feet to pass through the room in an hour.

The relative position and arrangement of the inlets and outlets regulate both the comfort and efficiency of the ventilation.

It is therefore necessary to consider what currents are caused in a room by ventilating openings.

In the first place, the question should be considered apart from an open fireplace, and apart from the question of warming the inflowing air.

If a room has two outer walls on opposite sides, and if an opening be made in each wall, and if the wind blows against one of the walls, there will be an increase of pressure against that wall, and a diminution of pressure against the wall opposite; consequently air will be forced in through the inlet on the side against which the wind blows, and be extracted on the other side.

Barrack-rooms have been occasionally ventilated by hollow beams, carried across the rooms from one outside wall to the other, communicating with the open air at both ends, and also provided with openings into the room, but having a wooden partition placed across in the centre of the beam, so as to compel it to act both as an inlet and an outlet when the wind is blowing against either outer wall, as in Fig. 9.

The action becomes more efficient if the beam be dispensed with, and the openings in the opposite walls retained. The most convenient form for such openings is the Sherringham ventilator ; Figs. 10, 11.

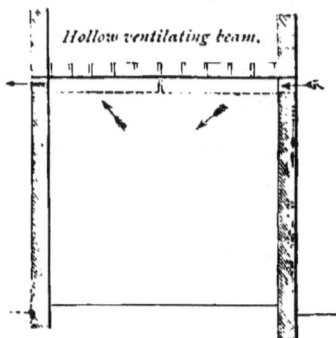 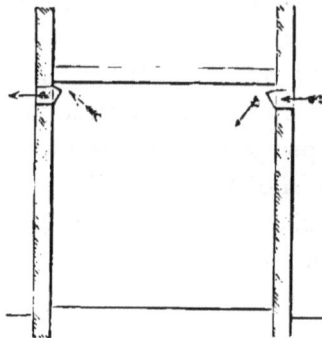

Fig. 9.   Fig. 10.

This consists of an iron air-brick or box inserted close to the ceiling of the room, and affording a direct communication with the external air. In order to prevent the air from coming in by stray currents, there is placed at the mouth of the opening within the room a hopper-shaped valve, hinged at its lower side and opening towards the ceiling ; the result of which arrangement is, that the inflowing current is thrown up towards the ceiling, and diffused to a greater or less extent in the general mass of air within the room.

Fig. 11.

This ventilator may, under certain conditions, act as an outlet ; but when the room is shut up, it would, especially with a fire in the grate, act as an inlet for fresh air.

Considered as an inlet, its principle and position are both

good, but acting by itself it is not a sufficient ventilator for rooms with a large number of occupants.

Inlets have been formed by vertical tubes, the opening to the outer air being made near the floor, and the tube being employed as a means of carrying the point at which the air is allowed to enter the room to a height of 5 or 6 feet or more above the floor. This is convenient in cases where necessities of construction make it desirable to place the opening to the outer air low down as compared with the point of entrance of the air into the room. The tube has the advantage of directing the inflowing current upwards towards the ceiling—but in consequence of the friction of the sides of the tube the velocity with which the air enters the room is less than it would be if it came in through a shorter channel and a more direct inlet inclined upwards like a Sherringham ventilator. The main objection to these tubes is that they form very convenient receptacles for dirt, insects, cobwebs, and dust, which after a time may injuriously affect the air passing through them. Moreover, inlets of this shape do not readily lend themselves to act the part of outlets when occasion requires, which is so convenient a feature of the Sherringham ventilator. Upon the whole Sherringham's is the most convenient form, and it is easily cleaned.

Fig. 12.

Where a room has two outside walls, and is provided with openings on both sides, this inflow and outflow of air is almost certain to go on continuously, in consequence of the movement of the outer air.

Where rooms have only one outer wall, other conditions prevail. The Sherringham ventilator, in such cases, would be found to act on occasions both as an outlet and an inlet ;

but its action would not be sufficiently energetic, consequently an additional outlet becomes necessary. This is best provided by means of a vertical shaft or tube carried from near the ceiling to above the roof. In a room possessing a chimney-shaft, and with few occupants, an Arnott or Boyle ventilator may usefully assist ventilation on those occasions where the fire is low or not lighted. It consists of an oblong metal frame inserted into the room chimney near the ceiling. Its object is to take advantage of the upward draught of the chimney in drawing the upper strata of the air of the room through the frame into the flue, while to prevent down-draughts of smoke into the room, a light silk or talc flap-valve, supported behind a perforated metal plate, is placed in the opening of the box into the room. This valve, like every other, requires certain conditions for its action. If the throat of the chimney be very wide, the quantity of air and smoke which pass up the shaft from below will be more than the chimney can accommodate at the narrower part, where the ventilator is placed, and smoke will consequently pass through the valve into the room. Wherever, therefore, valves of this sort are used, the throat of the chimney between the fireplace and the valve must be contracted to such an extent as to leave a balance in the draught to be supplied by air passing through the valve. As, however, the amount of this balance—in other words, the number of cubic feet of air which can pass through the valve into the chimney per hour—is very limited, this form of ventilator is not adapted to be the only outlet for a barrack-room, schoolroom, or for any room with several people in it.

Under such circumstances, as well as in rooms where this is not available, an outlet should be provided by a shaft carried from near the ceiling to above the roof. (Fig. 13.)

The velocity in these shafts, when acting freely of themselves, is dependent partly on the difference of temperature between the air in the room and the air without, but mainly on the amount of movement in the outer atmosphere, and partly on the adequate supply of fresh air in place of that removed, as well as on other circumstances. When the temperature is nearly equal, the upward draught is the result of movements in the outer atmosphere. It may be assumed that with a shaft a current varying from 2½ to 5 feet per second may be relied on, and often a stronger current will be found to prevail.

It will, however, sometimes happen, especially in rooms with a very short shaft, as, for instance, in rooms near the roof, that from the action of the wind the current becomes irregular, and is occasionally reversed, and produces down-draughts —so that the shaft becomes temporarily an inlet, and the Sherringham ventilator an outlet.

A reverse current in the shaft may be occasioned also if there is a large open fire in the room in-

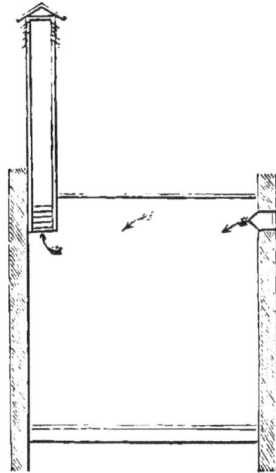

Fig. 13.

adequately supplied with fresh air ; or if a large and lofty hall or staircase is in proximity to the room—these may act like large shafts to draw in the air from the room and down the ventilating shafts.

Such an effect equally provides the room with fresh air ; but to prevent inconvenience to the occupants, it is advisable to terminate the shaft a little below the ceiling with a solid bottom and with side openings covered by inverted louvres, so that any down current would be directed towards the ceiling.

The sizes which were adopted for outlets and inlets of this description in the case of barrack-rooms are, for shafts or tubes in rooms next the roof, a sectional area of one inch to every 50 cubic feet of room-space ; for the floors next below the upper floor a sectional area of one inch to 55 cubic feet of room-space ; and, where the barrack consists of three floors, for the lower floors a sectional area of one inch to 60 cubic feet of room-space. For inlets one square inch for every 60 cubic feet of content of the room as the clear inlet area exclusive of gratings. This is the opening to the outer air—the opening into the room should afford an area larger than this, so as to reduce the velocity of the inflowing current. As the velocity of outflow and consequent inflow depends on temperature, provision must be made for closing either wholly or partially the inlets in cold weather ; and it is assumed that in very mild weather the sluggishness of the ventilation would be assisted by opening the windows. In barrack-rooms, whatever be the weather, the windows are opened during a certain time every day. These sizes were adopted on the basis of the cubic space allotted being 600 cubic feet per man, and that a volume of fresh air not less than 1200 cubic feet per occupant per hour should be admitted. The contamination of the air varies with the number of occupants. In hospital wards, in which the volume of air to be removed per occupant is greater than in barracks, the same sizes were still adopted ; because in hospitals, the cubic space per occupant being much greater, the change of air per occupant would be correspondingly greater. But in places where the cubic space is much less than this, such as in some schools, workhouse dormitories, or in common lodging-houses, a larger proportional area to cubic space may be found advisable.

There are forms of shafts which have been devised to act both as inlets and outlets, but in order to do so they require fixed conditions. Alter these fixed conditions and any of them may become wholly outlet or wholly inlet. The condition essential to their operation is, that the room to which they are applied be closed, and in a closed room their action is singular. If a number of people be crowded into a room with the fireplace closed and the doors and windows shut, and if a tube of an apparently sufficient area to afford ventilation for the inmates be carried from the ceiling of the room above the roof of the building, there will be an irregular effort at effecting an interchange between the air of the room and the outside air. The outer air will descend, and the inner air will ascend, in fitful, variable, irregular currents, and the room will be badly ventilated, if ventilated at all.

But singularly enough, no sooner is the tube divided longitudinally from top to bottom by means of any division, however thin, than its action becomes immediately changed—a current of air descends into the room continuously on one side of the partition, and a current of foul air ascends from the

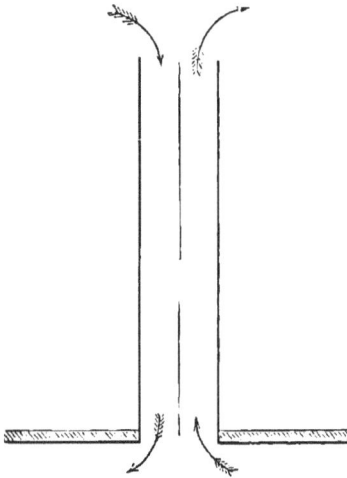

Fig. 14.

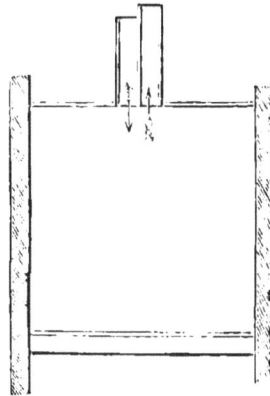

Fig. 15.

room continuously on the other side of the partition. One half of the tube supplies fresh air to the inmates of the room, and the other half removes foul air, so that if the size be properly adjusted the air in the room is kept sweet. (Fig. 14.)

Watson's Ventilator (Fig. 15) supplies this principle in its elementary form. It consists of a square tube with a division down the centre, one side affording a tube slightly higher than the other: it has no means of diffusing the descending current.

Mackinnel's ventilator (Fig. 16) professes to be an improved application of the same principle. It consists of two tubes, one within the other, leaving a space between them. The inner tube is the longer, and projects above the outer tube at its upper end ; the inner tube also projects a little below the opening of the outer tube in the ceiling, to give support to a circular flange projecting parallel with the ceiling, and concealing the opening of the outer tube. The action of this

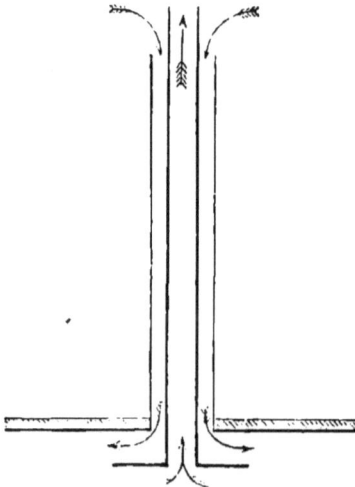

Fig. 16.

Plan

Fig. 17.

contrivance is as follows :—The 'greater length of the inner tube determines the upward current to take place in it ; it therefore becomes the foul air shaft. The outer tube becomes the fresh air inlet, and the descending current striking against the flange, is thrown out in the plane of the ceiling, and so diffused.

Muir's ventilator (Fig. 17) consists of a square tube, like

Watson's, divided into four parts, *A*, *A*, *B*, *B*, by partitions, inserted diagonally. These partitions are carried above the top of the tube, and the box is completed outside and above the roof by louvres instead of solid sides. The object of this arrangement of divisions and louvres is to secure not only upward and downward currents at ordinary times, but to take advantage of any movement of the external air, light, winds, &c., which, by striking through the louvres at any angle, would cause a stream of air to be projected down into the room, and would assist the extraction of the air on the side away from the wind.

All these ventilators act as desired in a closed room, but as soon as a door or window is opened they become simply upcast shafts ; they cease to supply air, and the air supply comes in from the other openings. Again, if there be a fireplace in the room, with a strong fire in it, and the doors and windows shut, the fire will supply itself from these ventilators, and they will become inlets.

It is obvious that these plans possess certain advantages in cases where they are applicable. In single rooms standing apart, such as churches, chapels, schools, libraries, &c., warmed by stoves or hot water pipes, and where the doors are kept shut for hours at a time, any of them will answer as ventilators.

In stables, also, of a certain construction, they will be more or less applicable.

In the case of living-rooms in houses they would not be applicable, both on account of the difficulty and cost of introducing the apparatus into a number of detached rooms on different floors, and on account of the existence of open fireplaces.

Inlets and outlets may be varied in form, but those mentioned are amongst the simplest forms. It must be borne in mind that so long as efficient action without draughts can be obtained, the best form to adopt is that which is simplest and most easily cleaned.

CHAPTER X.

It will have been apparent in the preceding chapters that the question of ventilation cannot be separated from that of the temperature of air. In a cold climate the air required to replace vitiated air must be warmed. In a hot climate cool air must be supplied. The question must be considered in the aspects of health, comfort, and economy.

Comfort and health are practically synonymous; and for these purposes in this climate the day-temperature of a room should be maintained at something between 58° and 66°. The night-temperature should not be so high, but it is not desirable that the night-temperature should fall below 40°. It should, however, be noted, that with a high temperature the quantity of oxygen present in the air is diminished. Thus, a cubic foot of dry air at 32° weighs 566.85 grains, and if the proportion of nitrogen and oxygen be assumed to be by weight 77 and 23 per cent., and the slight amount of carbonic acid be neglected, there will be in a cubic foot—

> 436.475 grains of nitrogen.
> 130.375   „   oxygen.
> ————
> 566.850

As a man draws, on an average, when tranquil, 16.6 cubic feet per hour into his lungs, he will thus receive $130.375 \times 16.6 = 2164.2$ grains of oxygen per hour.

At a temperature of 80° the foot of air weighs 516·38 grains, and is made up by weight of—

> 397·61 grains of nitrogen.
> 118·77 ., oxygen.
> ――――――
> 516·38

Therefore, in an hour, if a man withdraws 16·6 cubic feet, he will receive 118·77 × 16·6 = 1971·6 grains of oxygen per hour. Or, in other words, in an hour he would receive 192·6 grains of oxygen less with the higher temperature; that is to say, he would inhale an amount of oxygen equal to about 90 per cent. of the amount he would breathe at the lower temperature.

If saturated with moisture, a cubic foot of air will contain 130 grains of oxygen at 32°, and 112 grains at 100°.

The application of heat in an economical manner requires a consideration of the conditions under which heat is generated and diffused.

The standard unit of heat is the amount required to raise one pound avoirdupois of water at 32° one degree Fahrenheit. The specific heat of a body is the number of units of heat necessary to raise that body 1° Fahrenheit. The number of units of heat necessary to raise the temperature of one pound avoirdupois of the following bodies 1° Fahrenheit is—

		Unit of heat.			Unit of heat.
Air { constant volume	. .	·169	Marble and Chalk	. . . . .	·21
{ constant pressure	. .	·238	Glass	. . . .	·19
Water at 32° . .	. . .	1·00	Cast iron . . .	. . . .	·13
Pine-wood . .	. . . .	·65	Wrought iron . .	. . . .	·11
Oak	. . . .	·57	Copper	. . . .	·09
Ice	. . . .	·50	Lead	. . . .	·03
Olive oil . . .	. . .	·31			

It takes 966·1 British units of heat to evaporate 1 lb. of water under one atmosphere.

In order to find the quantity of heat required to produce a given rise of temperature in a given weight of a given sub-

H

stance, it is necessary to multiply together the rise of temperature, the weight, and the specific heat of the substance.

But the specific heat of bodies is somewhat greater at high
temperatures than at low temperatures. Thus if the specific
heat of water at 32° be taken as 1, its specific heat at 446°
Fahrenheit will be 1·05. 1 lb. of air at 32° contains 12·38 cubic
feet, and 1 cubic foot of air at 32° weighs ·08 lb. At 62° its
volume would be increased to 1·061, and its weight would be
·076 lb., whilst at 104° the volume would be 1·146 and the
weight ·07 lb., and at 212° the volume would be 1·365 and the
weight ·059 lb. All gases, dry air, and vapours out of contact
with their generating fluids, expand very nearly alike, and the
following formula is a fair expression of the rule of expansion—

$$m = M \times \frac{458·4 + t}{458·4 + T},$$

in which M = volume of gas or air, at temperature T, and m =
volume of air at new temperature t.

The rate of expansion increases slightly with the pressure.
When the pressure is not constant, the volume of any gas varies
in the inverse ratio of the pressure—the temperature remaining nearly constant. The pressure is the total pressure above
a vacuum. Thus one cubic foot of air has the pressure of
the atmosphere of 14·7, or say about 15 lb. per square inch
upon it to begin with, and thus its volume at an additional

pressure P above atmospheric pressure would be $\dfrac{1 \times 15}{15 + P}$.

Where there is a change both in temperature and pressure,
and P = pressure corresponding to temperature T and to
volume M, and p = pressure corresponding to temperature t
and volume m, the rule becomes—

$$m = M \times \frac{P}{p} \times \frac{458·4 + t}{458·4 + T}.$$

When water is present the rule becomes

$$m = M \times \frac{p + f}{P + F} \times \frac{458·4 + t}{458·4 + T};$$

in which M, P, T, and F are the volume, pressure, temperature, and elastic force of vapour corresponding to each other
in one case, and m, p, t, f in the other case.

The following table shows approximately the cold produced
by dilation and the heat produced by compression of air. The
volume at atmospheric pressure at the sea level and at 60°
Fahrenheit being 1·0.

Atmospheres.	No. of Inches of Mercury.	Volume of the Air.	Actual Temperature of the Air during the process. (Fahrenheit.)	Difference due to compression or expansion. (Fahrenheit.)
0·5	15	1·634	− 36°	− 96°
0·83	25	1·137	+ 33°	− 27°
1·000	30	1·000	+ 60°	0°
1·25	37·5	0·85	+ 94°	+ 34°
1·5	45	0·75	+ 124°	+ 64°
2·0	60	0·61	+ 175°	+ 115°

Thus if air at a temperature of 60° be compressed to half
its volume, or to a pressure of two atmospheres, its temperature
is raised to 175° Fahrenheit. If, on the other hand. it be
expanded so as to occupy double the space, the temperature
would be decreased to nearly 36° Fahrenheit below zero ; and
hence if air be compressed, and if the compressed air be
allowed to cool down to the temperature of normal air, and
the air be then allowed to expand, a degree of cold will be produced equal to the difference between that caused by the compression and the normal temperature. This affords an efficient
means of supplying cooled air for ventilating purposes.

The heating power of different kinds of fuel varies considerably, depending chiefly upon the proportion of carbon and
hydrogen which they contain. One atom or 12·5 lb. of
hydrogen combines with one atom or 100 lb. of oxygen to
form 112·5 lb. of water. The heat evolved by this combination is 62,535 units per pound of hydrogen. One atom or
75 lb. of carbon unites with 2 atoms or 200 lb. of oxygen
to form 275 lb. of carbonic acid. The heat evolved in this

process, according to Dulong's experiments, is 12,906 units per lb. of carbon. One atom or 75 lb. of carbon combining with one atom or 100 lb. of oxygen forms 175 lb. of carbonic oxide. The heat evolved in the process, according to Dulong's experiments, is 2495 units per lb. of carbon, and, according to the same experimenter, 1 lb. of carbon in the form of carbonic oxide burning to carbonic acid evolves 10,400 units of heat.

If the constituents of fuel be classed under C = carbon, H = hydrogen, and O = oxygen, and refuse, the theoretical evaporative power or total heat of combustion in units of evaporation per unit of weight of fuel from chemical analysis may be briefly expressed by the formula—

$$\text{Theoretical evaporative power} = 15C + 64\left(H - \frac{O}{8}\right)$$

as shown in the following table of examples of theoretical evaporative powers of fuel [1].

Carbon		15
Hydrogen		64
Various hydrocarbons	from	20 to 22½
Charcoal and coke	„	12 to 14
Coal, best qualities :—anthracite		15
„ „ bituminous	from	14 to 16
„ „ oxygenous	about	13½
„ „ brown	„	12
Peat, absolutely dry	„	10
Wood „	„	7½

The oxygen required for combustion is supplied by air.

[1] The following are the results which have been obtained by direct experiment by various experimenters of the value of several of the above substances as generators of heat.

	Units of heat per lb. of fuel.
Hydrogen burning to water will give	62,535
Petroleum will give	20,240
Carbon burnt to carbonic acid	12,906 to 14,040
„ „ oxide	2,495
Coal (mean of 97 varieties)	13,007
Charcoal from wood	12,000
Coke	10,970
Wood, perfectly dry	6,480
Wood, in ordinary state of dryness (i.e. with 20% of water)	5,040
Peat (dried naturally)	7,150
Peat (dried artificially)	8,736

One atom or 100 lb. of oxygen combines with 2 atoms or 350 lb. of nitrogen to form 450 lb. of air, equivalent to about 5800 cubic feet of air at 62° Fahrenheit. One pound of hydrogen requires for its combustion 8 lb. of oxygen, equivalent to 36 lb. or about 470 cubic feet of air; and one pound of carbon requires for its combustion 2·69 lb. of oxygen, equivalent to 12 lb. or 157 cubic feet of air at 62° Fahrenheit. The net weight of air which is chemically necessary for the complete combustion of a unit of weight of fuel may be expressed by the formula—

$$12\,C + 36\left(H - \frac{O}{8}\right),$$

where C stands for carbon, H for hydrogen, O for oxygen.

Nitrogen passes through a fire without material alteration, and for purposes of combustion the oxygen alone is available. Experience shows that the air which has passed through a fire retains a considerable proportion of its normal quantity of oxygen, and therefore for practical purposes of combustion the supply of air should be increased beyond the quantity which theoretical considerations show to be necessary, in the ratio of $1\frac{1}{2}$ to 1 or 2 to 1.

According to this view, 1 lb. of coal or charcoal requires for its combustion about 300 cubic feet of air at 62° Fahrenheit : 1 lb. of dry wood requires about 160 cubic feet of air. The efficiency of a furnace may be diminished by from ·2 to ·5 of its value through unskilful firing.

In a close stove perfect combustion depends on the area of the grate, and other apertures for admitting air, with reference to the fuel used, to the height of chimney or other means for drawing or propelling the air through the fuel, and to the power of regulating the inflow of air by dampers or doors. In open fireplaces, whilst a blazing fire will be best obtained by a grating under the fire, yet if the air be properly guided to the back as well as to the front and sides of the upper part of the fuel in an open fire, a bright fire will be obtained ; as for instance, with Dr. Arnott's plan of a closed bottom to the fireplace. In all cases when the supply of air is insufficient, carbonic oxide is formed.

When carbonic oxide is in course of formation in an ordinary grate burning coal, the fact is made known by the blue-coloured flame sometimes seen. When this occurs it is an evidence that there is an insufficient supply of oxygen at that part of the fire for supporting combustion The small number of units of heat evolved in the formation of carbonic oxide shows the wasteful effect of burning coal with an insufficient supply of air.

The effect of water in a combustible is to diminish the actual weight available for producing heat, as well as to absorb such a proportion of the heat generated as may be necessary for evaporating the water; therefore fuel should be as dry as possible. For instance, a fire supplied with damp wood will give out scarcely any heat.

When oxygen and carbon combine, the volume of carbonic acid formed is nearly the same as that of the oxygen consumed; therefore when a combustible contains carbon only, the volume of gas in the chimney is the same as that of the air entering the fire expanded to the volume due to the increased temperature.

The air will be heated to a temperature varying with the volume of air admitted, and with the heating power of the fuel. If half the air admitted be consumed, the temperature of the air as it leaves a coal fire will be 2256° Fahrenheit; but if only one quarter of the air were consumed the temperature would be 1159°. These temperatures are rapidly lost by radiation, and the temperature in the chimney of a well constructed steam boiler is not above 550°. Nor is it desirable that a greater temperature should be obtained, because beyond that temperature the increase of the volume of air by expansion limits the discharging power of the chimney.

Thus, assuming a chimney of 32 feet height, with an outside

temperature of 62°, and a temperature of air in the chimney of 192°, the velocity of the expanded hot air at exit is 24·69 feet ; whilst that of the cold air entering the chimney is 19·73 feet per second.

At a temperature of 582° in the chimney, the velocity of the expanded hot air at exit would be 52·6 feet ; whilst that of the cold air entering the chimney would be 26·3 feet ; and at 712° in the chimney the velocity of the expanded hot air at exit would be 58·44 feet per second, but in consequence of the expansion of the air, the velocity of the cold air entering the chimney would be only 25·99 feet per second, and higher temperatures would show a greater diminution.

Radiant heat passes through moderate thicknesses of air without sensibly heating the air, so that it may be assumed that air cannot be heated by the radiant heat from a flame or an open fire ; but the radiant heat warms the bodies which intercept the rays, and these bodies warm the air. For moderate temperatures the emission of heat from bodies by radiation is proportional to the difference of temperature, but for high temperatures and great differences of temperature, the proportionate emission is much greater. The radiant power of a body is equal to its absorbing power, and varies with the nature of the surface. The units of heat emitted or absorbed from a square foot of surface per hour, for a difference of 1° Fahrenheit are—

Silver, polished	·0265
Lead, sheet	·1328
Iron, ordinary sheet	·5662
Glass	·5948
Cast iron, new	·6480
Building stone, Plaster, Wood, Brick . .	·7358
Sand, fine	·7400
Water	1·0853

Heat is emitted and absorbed in an accelerating ratio in proportion as the difference of temperature increases between the body from which the heat is radiated and the body which receives the heat ; and with the same difference of temperature between the recipient and the radiant, the effect of the radiant will be greater according to the increased temperature of the recipient. In other words, the ratio of the emission of heat increases with the temperature.

Excess of temperature of the radiant over that of the recipient in degrees Fahrenheit.	Temperature of recipient of radiant heat in degrees Fahrenheit.			
	$32°$	$50°$	$104°$	$212°$
	Ratio of heat emitted in units of heat.			
$18°$	·997	1·075	1·355	2·150
$108°$	1·212	1·307	1·648	2·615
$324°$	2·07	2·23	2·81	4·46

Mr. Anderson made some experiments in 1875 on hot water pipes at low temperatures[1], and found that with a constant difference of temperature of $50°$ Fahrenheit, between the surface of the pipes and the air—

With the temperature of air =	$32°$	$39°$	$46°$	$53°$	$60°$
The total heat units given out by pipes was =	68.87	69.89	70.94	72.01	73.13

For the several reasons above mentioned, it is more economical to effect the warming of a given space by means of a highly heated surface than by a surface emitting a lower temperature.

[1] The total amount of heat radiated per square foot per hour by heated iron pipes may be found by the expression $m \times a^\theta$ $a^t - 1$), and the heat carried off by convection or contact with air in atmospheric air by the formula—

$$0·0372\left(\frac{p}{720}\right)^{0.45} \times t^{1.233},$$

in which $a = 1·0077$, θ = temperature of air surrounding the pipes, t = difference of temperature between the air and the pipes, both in centigrade degrees, p = height of barometer in millimètres, and m a coefficient of radiation.

Reduced to British units and 30 inches of the barometer, the formula becomes u = total units emitted per square foot per hour by radiation and convection

$$= m \times 1·00427^\theta \ (1·00427^t - 1) + 0·2853 \times t \ 1·233.$$

Péclet found the value of $m = 124·76$ for iron.

Mr. Anderson's experiments on hot water pipes gave $m = 122$ for cast-iron pipes, and $m = 250$ for wrought-iron pipes. Proc. Inst. C. E. vol. xlviii.

An open fire warms the air in a room by first warming the walls, floor, ceiling, and articles in the room, and these in their turn warm the air. Therefore in a room with an open fire, the air of the room is, as a rule, less heated than the walls. In this case the warming of the air depends on the capacity of the surfaces to absorb or emit heat ; except that the heat received by the walls may be divided into two parts, one part heating the air in contact with the wall, and the other passing through the wall to the outer surface, where it is finally dissipated and wasted.

In a close stove heated to a moderate temperature, the heat, as it passes from the fire, warms the surface of the materials which enclose and are in contact with the fire and with the heated gases, and transfers the heat through the materials to the outer surface in contact with the air ; and the air is warmed by the agency of this outer surface. If heated to high temperatures it gives out radiant heat, which passes through the air and warms the objects on which the rays impinge.

With hot-water pipes, the heat from the water heats the inner surface of the pipe, and this surface transfers its heat to the outer surface through the material of the pipes. The rate at which the heat can pass from the inner to the outer surface, and be thus utilised instead of passing away straight into the chimney, depends on the heat evolved by the fire, on the extent of surfaces exposed to the heat, on their capacity to absorb and emit heat, and on the quality of the material between the inner and outer surface as a good or bad conductor of heat.

The passage of heat through a body by conduction varies directly with the quality of material and with the difference between the temperature of the inner surface exposed to the heat and the outer surface exposed to a cooling influence, and inversely as the thickness between the surfaces.

If E = loss by conduction,

$\quad C$ = conducting power of the material,

$\quad t'$ = temperature of heated surface,

$\quad t$ = temperature of cooler surface,

$\quad G$ = thickness of material,

$$\text{then} \quad E = \frac{C(t'-t)}{G}.$$

The following is the quantity of heat in units transmitted per square foot per hour by a plate one inch in thickness, with a difference of 1° Fahrenheit between the temperature of the two sides of the plate :—

Copper .	515
Cast iron .	233
Wrought iron Plate	164
Wrought iron across the fibres	163
Lead	113
Stone (calcareous) .	13·68
Glass .	6·6
Brickwork	4·83
Plaster .	3·86
Wood, fir parallel to the fibres .	1·37
,, fir perpendicular to the fibres	·75
Wool .	·32

The conductivity of wrought iron is greater along than across the fibre. Other things being equal, copper is a better material than iron for conveying the heat from the fire to water or air; and coverings of brickwork, wood, or woollen fabrics are better adapted than iron for retaining the heat. The property which appears more than any other to make materials good non-conductors of heat is their porosity to air, and the fact that they retain air in their pores ; air being a very bad conductor of heat.

The direct warming of the air may be effected by stoves with brick or iron flues, or by hot-water or steam pipes. The sizes of the heating surfaces for this object must be proportioned to the volume of air required and to the degree of heat to be maintained; and it should also bear a proportion to the thickness and the capacity of the material for absorbing and radiating heat and for transferring heat from one surface to the other. When a large volume of air is supplied and removed for ventilation, rapidity in transferring the heat from the fuel to the air is an important consideration.

Brick stoves and flues are worse conductors of heat than iron stoves or flues; therefore the heat generated in a brick stove passes more slowly to the outer surface; but the surface of a brick stove parts with the heat which reaches it, that is to say it radiates the heat, somewhat more rapidly than do the surfaces of an iron stove. The slow conducting power of the material, and the greater thickness, of a brick stove, prevent alternations which may take place in the fire from being felt so much as with iron stoves or flues; and therefore the brick stove warms the air more equably without sudden variations; the air so warmed is free from objectionable effects; and where they can be conveniently applied, it is advisable to use brick stoves for warming air for ventilating purposes.

With an iron flue-pipe from a stove, almost the whole heat which any fuel is capable of developing may be utilised by using a long flue-pipe, horizontal for the greater part of its length, to convey the products of combustion to the outer air. The heat given out by a stove-pipe varies with the temperature from end to end, being of course greatest at the end next the stove, where the loss of heat is very rapid; and the amount of heat given out per square foot will vary at each point as the distance from the stove increases; and thus in dividing the pipe into lengths, each giving out an equal amount of heat, the length of the portion of pipe at the end further from the stove would be considerable, whilst the length required to give out an equal amount of heat near the stove would be very short. But not only does the amount of heat given out vary greatly from end to end of the flue-pipe, but the proportions into which the heat divides itself between the walls and the air vary greatly with the temperature.

Thus, with a stove-pipe heated at the end nearest the stove to a dull red heat of 1230° Fahr., and of sufficient length to allow the heat to be diminished to 150° at the further end, it would be found that at the stove end of the flue-pipe 92 per cent. of the total heat emitted by the pipe is given out by radiation to the walls, and only 8 per cent. to the air ; but at the exit end the heat is nearly equally divided, the walls receiving 55 and the air 45 per cent. Taking the whole length of such a pipe, the walls would receive 74 per cent. and the air 26 per cent. of the heat emitted. But with a flue-pipe heated to lower temperatures the air might receive half the heat or even more.

When therefore the object is to heat the walls rather than the air, which is sometimes the case, the temperature of the pipes should be high, and for this purpose stove-pipes are more effective than hot-water pipes, or low-pressure steam pipes. Hot-water pipes heated under pressure on Perkins' system, and high-pressure steam pipes would radiate a proportion of heat to warm the walls.

At high temperatures there will be practically little difference of effect between horizontal and vertical flue-pipes, because the heat given out is principally that due to radiation, which is independent of the form and position of the radiant.

An adequate proportion of flue-pipes to the form and size of the stove involves a large surface for the flue-pipe ; and with a careful observance of proportion, as much as 94½ per cent. of the heat in the fuel has been utilised, only 5½ per cent. being carried away.

In deciding on the proper height of the flue-pipe as a chimney for a stove, it is necessary to consider the three cases of (1) a vertical flue-pipe and of a pipe with a uniform upward slope, (2) of a pipe with the part next the stove vertical while the rest is horizontal, and (3) of a pipe with the part next the stove horizontal and the vertical part at the exit end.

The mean temperature of the internal air in the flue regulates the discharge which. after allowing for friction, is due to the difference in vertical height between the stove end of the flue and the chimney end. Therefore, in the first case —viz. with a flue either vertical or with a uniform slope—the mean temperature of the flue may be taken. In the second case, the high temperature at the vertical part of the flue may render a shorter vertical flue equally efficient with the inclined flue ; but in the case where the horizontal part of the flue is nearest to the fire, and the vertical part the furthest removed from it, the temperature in the vertical part of the flue will be lower, being further from the stove, and the height of the vertical portion must be greater than that of the uniformly inclined flue.

There are however several serious objections to iron stoves, especially for small rooms. A long flue-pipe is unsightly, and on that account often inadmissible.

Iron stoves heat rapidly, and easily become red-hot, therefore the effect produced is unequal both on the air and on persons in the vicinity of the stove. An iron stove cools down with rapidity when the fire is low. The flue-pipe gives out unequal degrees of heat in the different parts of its length. With an iron stove the temperature at 18 inches from the stove has been found to exceed by 27° Fahrenheit that observed at 6 feet from it ; and with a red-hot stove the difference in that short distance was in some cases as much as 45°.

Carbonic oxide has been found in air heated by iron stoves. This can only occur provided the stove be very highly heated ; but a high temperature is a liability to which iron stoves are subject. Iron very highly heated may take up the oxygen from the carbonic acid prevalent in the air of an occupied room, and thus reduce the carbonic acid to the condition of carbonic oxide. Moreover, the quantity of dust in a room, which almost always contains organic matter, may under these conditions of temperature somewhat influence the presence of carbonic oxide.

General Morin alleges that he found these effects to be nearly three times as great with cast-iron as with wrought-iron stoves. It has also been alleged that the carbonic oxide generated in the fire may permeate through the cast iron of the stove, if very highly heated or of inferior porous metal, into the surrounding air. Carbonic oxide may also be produced by the oxygen of the air acting on the carbon in the cast iron if heated to a red heat. The effect would be diminished by the presence of moisture in the air. Consequently the use of vessels containing water on metal stoves has been recommended. Whatever may be the value of these allegations, the use of surfaces of iron heated to a red heat for warming air for ventilating purposes is seriously objectionable. Hence, whenever iron stoves or cockles are used for heating air, care should be taken to prevent the iron from attaining a high temperature, and with this object all iron stoves should have a lining of fire-brick, so as to prevent the fire from coming in direct contact with the iron : such an arrangement prevents these inconveniences, and preserves greater regularity in the heating of the air. An advantageous arrangement would be to provide the iron with an outer surface of glazed enamel, because the iron would convey the heat rapidly from the fire to the surface, while the enamel surface would radiate the heat more rapidly than the iron surface.

Hot-water pipes for warming air are free from many of the objections arising from the direct application of heat to iron, because the heat can be regulated with exactness.

Water boils at 212° Fahrenheit under the atmospheric pressure of 14·7 lb. per square inch, or 39 inches of mercury, i. e. at about the sea level. Under one-half that pressure, viz. 7·3 lb. per square inch, or 15 inches of mercury, it boils at 180°; and under a pressure of four atmospheres above the ordinary atmospheric pressure, or 44 lb. per square inch, it boils at 291°; and under a pressure of 10 atmospheres, or 132 lb. per square inch, it boils at 357° Fahrenheit. Thus a high temperature may be obtained from water without generating steam by heating it under pressure. Salt water, saturated with 41·2 lb. of salt per 100 lb. of water, boils at 227°, and freezes at about the zero of Fahrenheit.

Steam is generated under the mean atmospheric pressure of 30 inches of mercury when the boiling-point is attained, i. e. 212°; but before the water becomes vapour, such further amount of heat equal to 966° is absorbed and becomes latent as would suffice to raise the water to 1178° if it did not turn into steam. The temperature of 1178°, which is thus required to produce steam, is necessarily constant, and consequently a greater or less amount of heat becomes latent according as the pressure is below or above the atmospheric pressure. This latent heat is given out on the reconversion of steam into water.

Pipes may be heated either by hot water or by steam. The higher the temperature, the greater is their comparative effect on the warming of air. Thus pipes heated by hot water under pressure convey heat to the air with greater rapidity than pipes heated by hot water at low pressures ; and steam pipes are more effective than hot-water pipes ; and steam at a high pressure is more effective than low-pressure steam. One advantage of heating by steam or by water under considerable pressure is the high temperature obtained in the pipes, and the consequent radiation of a larger proportion of heat to the walls of a room, than takes place with pipes heated under ordinary pressures. When steam is employed to warm air for ventilating purposes it is often necessary to adopt special means to moderate the temperature in warm weather.

There is moreover some practical inconvenience attending the use of high-pressure steam in localities where an experienced workman is not near at hand. For this reason, hot-water pipes have been generally preferred for warming ordinary houses.

The efficient action of hot-water pipes depends upon the

upward flow of the heated and expanded water, as it passes
from the boiler, being made as direct as possible, and so protected as to lose little heat between the boiler and the place
where the heat is to be utilised. The return pipe, which
brings back the water to the boiler after it has been cooled
down by the abstraction of heat in warming the air, should be
passed in to the bottom of the boiler as directly, and in as
uniform a line from the place where the heat has been used,
as possible. The velocity of flow in the pipes will depend upon
the temperature at which the water leaves the boiler, the height
to which the heated water has to rise, and the temperature at
which it passes down the return pipe back into the boiler.
The efficiency of a hot-water apparatus will be regulated by
these conditions, by the sizes of the pipes, and by such other
conditions as affect the flow of water in pipes.

It will be evident that to obtain an equal velocity of flow
when the height of the vertical column is small, the temperature at which the water returns to the boiler must be lower
than when the vertical column is long. Therefore when the
boiler or source of heat is very near the level of the pipes
for heating the air, the average temperature which can be
obtained in the pipes will be lower than when the vertical
column is long. Hence, the heating surface of the boiler,
the area of the grate which regulates the flow of air to the
fuel, and the surface of pipe which enables the heat from
the boiler to be utilised, must be regulated with reference
to this difference of level.

It may further be assumed that with small pipes, the
temperature being constant, the velocity of flow in the pipe
necessary to furnish a given amount of heat will vary in
the ratio of the length of the pipe.

When the water circulates through the pipes by virtue of
the difference of temperature of the flow and return currents
only, it is impossible to count upon a greater mean temperature in the pipes than from 160° to 180°, because above that

temperature the water in the boiler begins to boil, and causes an overflow of the supply cistern and escape of steam at the air pipes. In order to obtain a sufficient velocity of circulation for long distances, or with small differences of level, a forced circulation may be resorted to, as has been done by Messrs. Easton and Anderson at the County Lunatic Asylum at Banstead.

There two pipes are laid side by side. one of which communicates with the boilers and is termed the flow-pipe; the other, termed the return-pipe, is connected with the feed-cistern for the boilers, which cistern is situated above the level of the boilers.

Both pipes are connected with the various coils to which the heated water is desired to be conveyed, by valves which can be opened or closed at will. An Archimedean screw pump is fixed on the return-pipe near the point where the pipe ascends into the cistern. This pump is always kept at work. When the communications between the flow and return pipes are closed, the screw simply slips through the water; as soon as any communication is opened, the screw draws the water along the pipe, and forces it into the cistern, thus ensuring a constant circulation.

By Perkin's system the water is heated under considerable pressure, and a higher temperature is thus obtainable than with ordinary pressures.

In its simplest form the apparatus consists of a continuous or endless iron tube of about one inch diameter, closed in all parts and filled with water. The joints are screw-joints connected within a socket forming a right and left hand screw. About one-sixth part of the tube is coiled in any suitable form and placed in the furnace, forming the heating surface, and the other five-sixths are heated by the circulation of the water which flows from the top of this coil; and after having been cooled in its progress through the building, returns to the bottom of the coil to be re-heated.

I

Water when heated from 39°1 to 212° Fahrenheit expands about 5 per cent. of its original bulk. Therefore, in order to provide for the increased volume of the water when heated, a tube called an expansion tube is placed above the highest level of the smaller tubes which convey the heat to the distant parts of the building.

This tube is of larger diameter than those used as heating surfaces, and its length and capacity are proportioned to the quantity of tube to which it is attached.

The filling tube of the apparatus is placed on a level with the bottom of this expansion tube so as perfectly to fill all the small tubes, and yet prevent the possibility of filling the expansion tube itself. The expansion tube being then left empty, allows the water as it becomes heated to expand without endangering the bursting of the smaller tubes.

The apparatus is filled by an opening connected with the lowest line of tubing, so that the water, as it rises, drives the air before it, and out through an opening in the expansion tube. Great care must be taken to expel all air from the pipes by repeatedly forcing water through them. When the pipes are filled, both the opening in the filling tubes and the opening in the expansion tube are closed by screw-plugs.

The form and size of the furnace varies according to the locality and the work the pipes have to do.

A temperature of as much as 300° Fahrenheit can be obtained in the tubes.

Steam in lieu of hot water is especially applicable, either where steam for other purposes is in use, as for instance where the exhaust steam from an engine is available, or where heating is required on a large scale.

Steam-heating may be either on what is termed the high-pressure system or the low-pressure system. The high-pressure system of steam-heating is generally considered to mean the system which allows the steam to escape after use or else to be passed into a feed water-tank, whence it has to be pumped back into the boiler.

The low-pressure system is considered to mean the system in which there is a flow-pipe from the top of the boiler for the steam, and a return-pipe into the bottom of the boiler for the water of condensation, so arranged as to require no pumping.

A low-pressure gravity system of steam-heating is probably the most economical and convenient form of steam-heating appliance, as it is equally applicable to heat a single room or a large building. Its principal merits, when well done, are : it is safe ; noiseless ; the temperature of the heating-surface is moderate and uniform ; all the water of condensation is returned into the boiler, except a very small loss from the air-valves ; it is easy to keep the stuffing-boxes of the heater-valves tight ; and it is no more trouble to manage than a hot-water apparatus.

With a high-pressure system the waste of heat may amount to as much as 30 per cent., with traps which discharge into an open tank or into the atmosphere.

Where exhaust steam is used, it may be assumed that the capacity of heating is almost in a ratio with the fuel expended under the boiler. There is some cooling in passing through the engine, and along the pipes to the rooms requiring to be heated, but this is slight if the pipes and cylinder are covered with a good non-conducting material, so that the condensed water may be taken back hot into the boiler.

So far as economy in heating is concerned, there is not much difference between the use of exhaust steam and of that taken direct from the boiler. For instance, if the steam is taken from the boiler direct to the pipes at five atmospheres, the temperature would be a little over 300°. If, on the other hand, a comparative capacity of steam were allowed to pass through the engine to create power, and discharged into the pipes at one atmosphere, it would decrease in temperature to something over 212°, but it would increase in bulk according to the expansion ; and by an increased heating surface a larger proportion of heat can be practically utilised than shown by the difference of the temperature.

As regards the extent to which exhaust steam can be carried, Mr. Boulton, who has perfected this system, mentions a case in which from an engine of 25-horse power nominal, with a 17-inch cylinder, the exhaust steam travelled about 200 yards in a direct line, and passed along various branches, amounting in the aggregate to about 2386 yards, or 1⅓ miles of 1½-inch pipes, and then into the tank to warm the feed water for the boiler.

The use of steam for heating on a large scale is exemplified in many of our principal buildings, and amongst others in the Houses of Parliament. An interesting application of steam for heating is that adopted at Lockport, a town in the United States, of which a description has been given by Mr. George Maw ; the object in this case being to avoid the necessity of having separate fires in each house.

Two hundred houses were heated from the central supply through about three miles of piping, radiating from the boiler-house ; this building contained two boilers each 16 ft. by 5 ft., and one boiler of about half the size. The steam during the winter was maintained at a pressure of 35 lbs. to the inch, with a consumption of four tons of anthracite coal in 24 hours.

This boiler pressure of 35 lbs. was maintained through the entire length of the three miles of main piping up to the points of consumption, at each of which there is a cut-off under the control of the consumers.

The first 600 feet of mains from the boilers are 4 inches in diameter. There are 1400 feet of 3-inch pipes ; 1500 ft. of 2½-inch pipes ; and 2000 feet of 2-inch pipes. The supply-pipes from these mains to the houses are 1½ inch in diameter, and within each house ¾-inch pipes are used. In addition to the cut-off tap from the main under the control of the consumer there is a pressure-valve regulated to a 5 lb. pressure under the control of the company, and beyond this is an ingeniously constructed meter, which not only indicates the total consumption in cubic feet of steam, but also the quantity of steam used in each apartment. At each 100 feet of main an expansion valve like an ordinary piston and socket is inserted, allowing an expansion in each section of 100 ft. of $1\frac{3}{4}$ inch for the heat at 35 lbs. pressure. No condensation occurs in the mains. They are covered with a thin layer of asbestos paper next the iron, then a wrapping of Russian felt, and are finally wrapped round with Manilla paper like smooth light brown paper, and the whole encased in timber bored out three-quarters of an inch larger than the covered pipes, and laid along the streets like gas-pipes.

The distribution of heat in the apartments is by means of radiators, consisting of inch pipes 30 inches long placed vertically either in a circle or in a double row, and connected together top and bottom ; they are furnished with an outlet pipe for the condensed water which escapes at a temperature a little below boiling, sufficient in quantity for the domestic purposes of the house, or it may be used as accessory heating power for horticultural and other purposes.

The steam was also applied at a distance of over half a mile from the boilers for motive power, and it was stated that two steam engines of 10-horse and 14-horse power were occasionally worked from the boilers at a distance of half a mile with but a slightly increased consumption of fuel.

The laid-on steam was used for cooking purposes, for boiling, and even baking ; in a house three-quarters of a mile from the boilers, a bucket of cold water could be raised to boiling heat in three minutes, by the passage of the steam through a perforated nozzle plunged into the bucket.

As in the case of gas supply, the Steam Supply Company lay their pipes up to the houses, the consumer paying for all internal pipes, fittings, and radiators.

In a moderately sized eight-roomed house the expenses of these amounted to 150 dollars, or a trifle over £30, and in larger houses with more expensive fittings to 500 dollars, or about £100.

The working expenses consist of but little more than the coal and the wages of two firemen.

The annexed diagram (Fig. 18), resulting from Mr. Anderson's experiments, is published in the Journal of the Institution of Civil Engineers for 1877, and shows the total units of heat

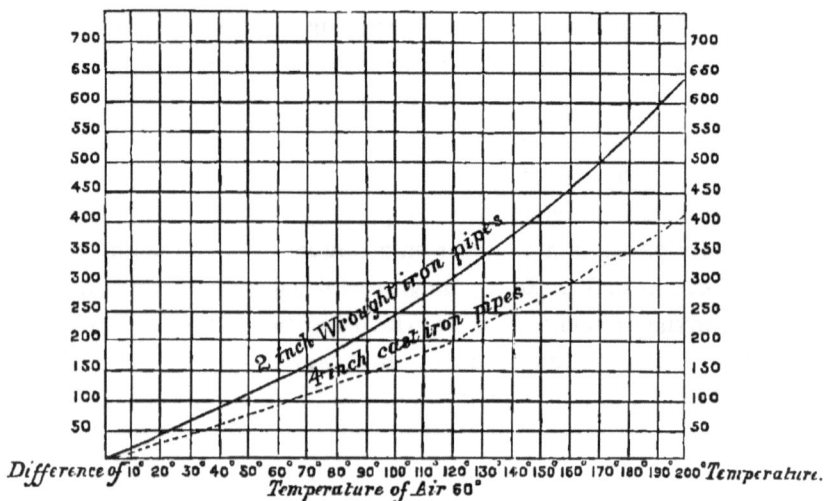

Difference of 10° 20° 30° 40° 50° 60° 70° 80° 90° 100° 110° 120° 130° 140° 150° 160° 170° 180° 190° 200° *Temperature.*
Temperature of Air 60°

Fig. 18.

given out by cast-iron and wrought-iron pipes per square foot of surface per hour for various differences of temperature, applicable either to hot-water or steam pipes.

Suppose, for example, it is required to know how much heat will be given out by 4-inch pipes at 190° in a room, the temperature of which is 60°, the difference of temperature being 130°: look along the line of abscissæ for 130, and the ordinate then gives 232.7 units for 4-inch pipes, and 356 units for 2-inch wrought-iron pipes per square foot per hour.

When the form of apparatus for heating has been decided

on, the amount of heating surface to be afforded for purposes
of ventilation and warming depends mainly upon the volume
of air to be admitted and removed, and the temperature
desired to be maintained ; but in any given building there
are other circumstances to be taken into account—viz. the
position, aspect, subsoil, temperature of locality, shape and
size of building, extent, position, and thickness of walls, size
and form of windows, skylight, and suchlike matters, which
affect either the temperature of the incoming air or the conditions which determine the loss of heat in a building.

A rough empirical rule is suggested by Mr. Anderson for
hot-water pipes in this country—viz. that 1 square foot of
heating surface is required for every 65 cubic feet of content
of the space to be warmed, and in a greenhouse 1 square foot
to every 24 cubic feet. This empirical rule does not however
appear to be based on the important sanitary considerations
of the renewal of air. A similar rough approximate rule
for pipes heated with exhaust steam has been given by
Mr. Boulton as follows: 1 superficial foot of steam pipe for
each 6 superficial feet of glass in the windows ; 1 superficial
foot of steam pipe for every 6 cubic feet of air removed for
ventilating purposes per minute, 1 superficial foot of steam
pipe for every 120 superficial feet of wall, roof, or ceiling,
allowing about 15 per cent. on the amount thus obtained for
contingencies. He states that, roughly speaking, the exhaust
steam due to one-horse power can be made to warm 30,000
cubic feet of space.

If the fresh air for ventilation is warmed in underground
chambers, such places should be so formed as to be dry and
well-drained, and the channels leading therefrom constructed
of such form and materials that the air shall lose as little
heat as possible in transmission.

When air is warmed, its capacity for moisture is increased ;
thus if air at 32° Fahrenheit be in a condition of complete
saturation from moisture, which is represented by 100 degrees,

and if the air be then raised to a temperature of 52°, it would have only 46 degrees of saturation, and appear to be dry; whereas if the temperature be raised to 72° Fahrenheit, the degree of saturation would be diminished to 23 degrees, and the air would appear to be extremely dry.

In a damp climate, up to a certain point, the drying of the air caused by raising the temperature is as much a source of comfort as the warmth. But when the outer air is dry, the additional dryness caused by warming the air may in some cases become inconvenient.

For reasons already mentioned, the open fire does not directly warm the air to so great an extent as stoves or heated pipes, therefore the dryness is generally not found to be objectionable with an open fire. But with stoves or hot-water pipes, it is frequently necessary to provide vessels of water on or near the stoves or heated pipes, to supply additional vapour to the air when required.

The conclusions which follow from these considerations are, that where a room is heated by warmed air passed through flues into the room, without any source of heat in the room itself, the air imparts its heat to the walls. The air is thus warmer than the walls. When a room is warmed by an open fire, on the other hand, the warming is effected by the radiant heat from the fire ; the rays from the fire pass through the air without sensibly warming it, the radiant heat warms the walls and furniture, and these impart their heat to the air. Therefore the walls in this case are warmer than the air. Consequently in two rooms, one warmed by an open fire and one by hot air, and with air at the same temperature in both rooms, the walls in the room heated by hot air would be some degrees colder than the air, and therefore colder than the walls of the room heated by an open fire, and these colder walls would therefore abstract heat from the occupants by radiation more rapidly than would be the case in the room heated by an open fire. And to bring the walls up to the same temperature in the former case—viz. in the room heated with hot air—the air of the room would require to be heated to an amount beyond that necessary for comfort, and therefore to a greater amount than is desirable on sanitary grounds, seeing that warmed air contains less oxygen than cooler air. As sick persons are more sensitive to such influences than persons in health, this would appear to be the reason why in hospital wards the warming by means of an open fire has been always recognised as preferable to warming by hot air.

To a certain extent, dependent on the temperature of the heating surface, a similar effect takes place in rooms warmed by close stoves, hot-water pipes, or low-pressure steam pipes; but in a room warmed by steam pipes heated to a high temperature, the effect on the walls would tend to approximate in some degree to that produced by the open fire in proportion as the temperature of the pipes is raised. Where the object is to warm the walls as well as the air, high-pressure steam is a more advantageous mode of heating than hot-water pipes. The arrangements for warming a building of any extent should be made under expert advice.

CHAPTER XI.

ONE pound of coal is far more than sufficient, if all the heat of combustion is utilised, to raise the temperature of a room, 20 feet square and 12 feet high, to 10 degrees above the temperature of the outer air. If the room were not ventilated at all, and the walls were composed of non-conducting materials, the consumption of fuel to maintain this temperature would be very small; but, in proportion to the change of the air of the room and to the escape of heat through the walls, windows, ceiling, &c., so would the consumption of fuel necessary to maintain that temperature increase. If the volume of air contained in the room above mentioned were changed every hour, one pound of coal additional would be required per hour to heat the inflowing air, so that to maintain the temperature at 10 degrees above that of the outer air during 12 hours would require 12 lbs. of coal.

The principle of the ordinary open fireplace is that the coal shall be placed in a grate, to which air is admitted from the bottom and sides to aid in the combustion of the coal; and an ordinary fireplace, for a room of 20 feet square and 12 feet high, will contain from about 15 to 20 lbs. at a time, and, if the fire be kept up for 12 hours, probably the consumption will be about 100 lbs., or the consumption may be assumed at about 8 lbs. of coal an hour. There is this drawback to the open fire, that whatever may be the theoretical perfection of the fireplace for efficient combustion, in practice it is impossible at all times to avoid smoke with an ordinary chimney. The only practical way of removing smoke would be to draw the fumes through a chamber where they could be washed so as to allow the cleansed gases only to be passed into the atmosphere.

As already mentioned, the radiant heat from the fire does not warm the air of the room ; the rays from the fire warm the sides and back and parts adjacent to the grate, they warm the walls, floors, ceiling, and furniture of the room, on which they impinge, and these impart heat to the air. The form and material of the fireplace can thus assist materially the warming of the air. The rays should impinge more freely on the walls and floor than on the ceiling. A projecting chimney-piece with a surface favourable to the absorption and emission of heat would be more favourable to the warming and circulation of the air than one which would allow the rays to pass to the ceiling. In an ordinary fireplace the sides should be splayed, as in the Rumford form of grate ; the sides and back should be of non-conducting material, with a surface favourable to the rapid absorption and emission of heat. Thus brick or tiles are better than iron for this purpose. Similarly, the degree to which the materials of the walls or floor of the room are unfavourable to conduction but favourable to the absorption and emission of heat will have a bearing on the capacity of the room for warmth.

One pound of coal may be assumed to require, for its perfect combustion, 160 cubic feet of atmospheric air ; 8 lbs. would require 1,280 cubic feet ; but at a very low computation of the velocity of the gases in an ordinary chimney-flue the air would pass up the chimney at a rate of from 4 to 6 feet per second, or from 14,000 to 20,000 cubic feet per hour ; with the chimneys in ordinary use, a velocity of from 10 to 15 feet per second often prevails, giving an outflow of air of from 35,000 to 40,000 cubic feet per hour. This air comes into the room cold, and when it is beginning to be warmed it is drawn away up the chimney, and its place filled by fresh cold air. A room 20 feet square and 12 feet high contains 4,800 cubic feet of space. In such a room, with a good fire, the air would be removed four or five times an hour with a moderate draught in the chimney, and six or eight times with a blazing fire. The atmosphere of the room is thus being cooled down rapidly by the continued influx of cold air to supply the place of the warmer air drawn up the chimney; and General Morin estimated that of the heat generated by fuel in an ordinary open fireplace about one-eighth only is utilised in the room. The very means adopted to heat the

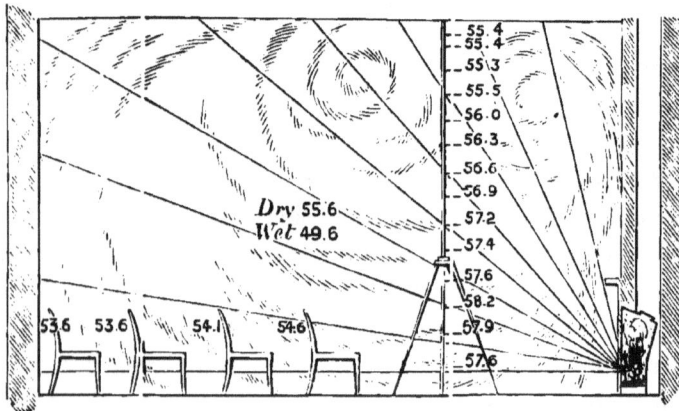

Dry 55.6
Wet 49.6

55.4
55.4
55.3
55.5
56.0
56.3
56.6
56.9
57.2
57.4
57.6
58.2
57.9
57.6

53.6 | 53.6   54.1   54.6

Fig. 19. Sketch of Experiment made by Mr. Campbell in 1857, showing the movement of currents of air in a room with an open fireplace.

room tends to produce draughts, because the stronger the direct radiation, or rather the brighter the flame in open fireplaces, the stronger must be the draught of the fire and the abstraction of heat.

A fireplace is thus powerful enough to draw into the room all the air it wants ; and for this purpose will use indiscriminately all other openings, whether inlets or outlets if necessary.

The only way to prevent draughts is to adopt means for providing fresh warmed air to supply the place of that removed.

The way in which an ordinary open fireplace acts to create

circulation of air in a room with closed doors and windows, is as follows :—The air is drawn along the floor towards the grate ; it is then warmed by the heat which pervades all objects near the fire, and part is carried up the chimney with the smoke, whilst the remainder, partly in consequence of the warmth it has acquired from the fire, and partly owing to the impetus created in its movement towards the fire, flows upwards towards the ceiling near the chimney-breast. It passes along the ceiling, and as it cools in its progress towards the opposite wall, descends to the floor, to be again drawn towards the fireplace.

It follows from this, that with an open fireplace in a room, the best position in which to deliver the fresh air required to take the place of that which has passed up the chimney, is above the projecting chimney-piece, and at any convenient point in the chimney-breast, between the chimney-piece and the top of the room, for the air thus falls into the warmer upward current, and mixes with the air of the room, without perceptible disturbance ; and it also follows that if two fire-places be placed in a room, they should, unless the room be very large, both be on the same side of the room, and if shafts are required for ventilating a room in addition to the open fireplace, they also should be placed on the same side of the room as the fireplace, but as far from it as possible.

The open fireplace thus presents special advantages for securing efficient ventilation by means of the circulation of air which it creates. It makes the room in which it is in use independent of other means of extraction of air, unless the room is very crowded, or beyond the size for which the fire-place is calculated.

The air thus extracted must be drawn into the room from somewhere, and unless arrangements be made for supplying the room with warmed fresh air, cold air finds its way into the room through the special inlets, if any are provided ; if not, through the chinks of the windows and doors, or wherever it

can get in most easily. The large volume of fresh air required
to supply that drawn up the chimney cannot always be
warmed with sufficient rapidity by contact with the walls and
furniture only; the temperature in different parts of the room
is therefore frequently very unequal, and the occupants are
subjected to draughts; and if there are two fireplaces in the
same room unprovided with special means for the admission
of fresh air, one of which is not lighted, the air is often drawn
down the vacant chimney.

It is therefore desirable in cold weather to replace some

Fig. 20. Section of a room with a ventilating grate and warm-air flue,
showing action of fire in producing circulation of air.

part at least of the air drawn up the chimney by air partially
warmed. This can be effected in various ways, either by a
service of hot water-pipes carried into all the rooms, to coils
placed in front of the fresh-air inlets, so as to warm the fresh
air as it enters the room; or by bringing warmed air into the
room by flues from a central heating apparatus. But the simplest
plan is to utilise some of the waste heat which in the case of
an ordinary open fireplace passes away unused up the chimney.

The ventilating fireplace was designed with the object of
obviating the above-named objections to the common fire-

place, of utilising spare heat, and of providing such adequate means of ventilating the soldiers' rooms in cold weather, when the windows are shut, as would not be liable to be deranged.

Fresh air is admitted to a chamber formed at the back of the grate, where it is moderately warmed by a large heating surface and then carried by a flue, adjacent to the chimney-flue, to the upper part of the room, where it flows into the currents which already exist in the room. These grates have never been patented, and therefore manufacturers do not care to suggest their use.

The body of the stove is of the best cast iron, and consists of three pieces, properly connected by screws. The first piece forms the moulded projecting frame; the second the body of the grate; and the third the nozzle or connexion with the smoke-flue, the bottom flange of which is bolted to the back of the grate. The stoves are of three sizes. The largest has an opening for fire of 1 ft. 9 in. wide, and was intended for rooms containing from 7,500 to 10,000 cubic feet; it weighs about 3 cwt. 1 qr. 10 lbs. The second, or medium size, has an opening for fire 1 ft. 5 in. wide and was intended for rooms containing from 3,600 to 7,500 cubic feet; it weighs about 2 cwt. 3 qr. 5 lbs. The third, or smallest size, has an opening for fire 1 ft. 3 in. wide, and was intended for rooms containing 3,600 cubic feet and under; it weighs about 2 cwt. 2 qrs.

The figures appended (Figs. 21–24) show an elevation, section, and plan of the second or medium-size stove, the extreme dimensions of which are 40 inches wide by 42 inches high. The projecting moulded frame enables the stove to be applied to any existing chimney-opening.

The fireplace has a lining of fire-lumps in five pieces—two sides, one back-piece, and two bottom pieces, moulded to the form shown in the woodcut. The object of this fire-clay lining or cradle is to prevent the contact of the incandescent fuel with the iron, and to preserve a high uniform temperature in the vicinity of the fuel to assist the combustion.

The bottom is partly solid, being made of two fire-lumps placed one on each side, and supporting an intermediate cast-iron fire-grating, which occupies about one-third of the bottom of the grate ; by this means, whilst the draught is checked by the solid part of the bottom of the grate, and the con-

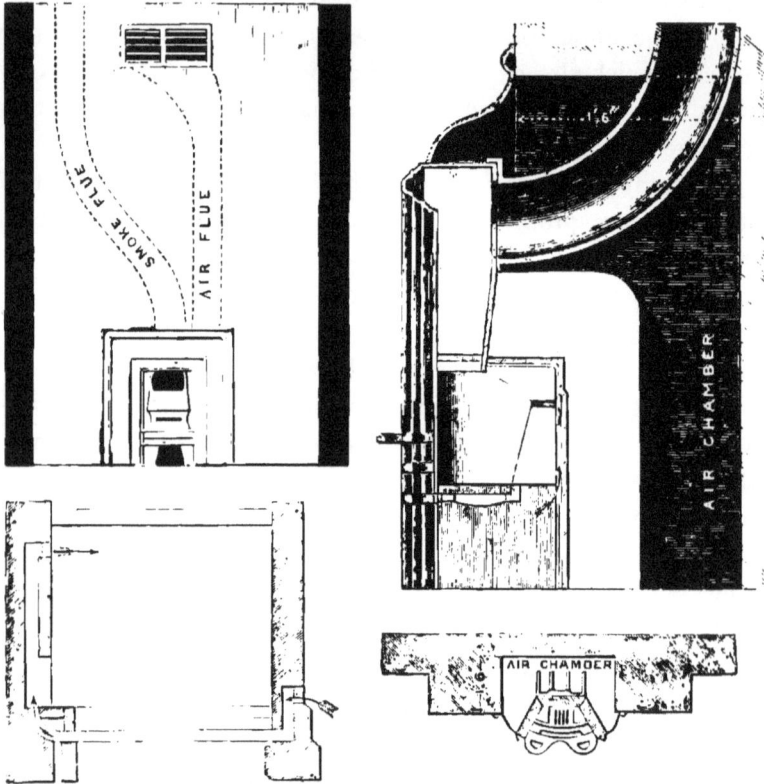

Figs. 21–24.

sumption of fuel reduced, a sufficient supply of air is obtained for combustion through the grating to secure a cheerful fire. A clear space, half an inch deep, is formed between the back piece of fire-lump and the iron back of the grate, through which a supply of air passes from the ash-pit under the grate, and through a slit in the fire-lump, on to the upper part of the back of the fire. The air thus brought into contact with heated coal is received at a high temperature, in consequence of passing through the heated fire-lump, and is forced into contact with the gases from the coal by means of the piece of fire-lump which projects over the fire at the back of the grate ; and thus a more perfect combustion of the fuel is effected than with an ordinary grate, and the creation of smoke is prevented : in fact, with care, almost perfect combustion of the fuel, and consequent utilisation of the heat, can be obtained.

Whilst the incandescent fuel and flame are kept away from actual contact with the iron back of the stove, the heated gases from combustion, and such small amount of smoke as exists, are compelled, by the form of the back of the grate, and the iron part of the smoke-flue, to impinge upon a large heating surface, so that as much heat as possible may be extracted from the gases before they pass into the chimney ; the heat thus extracted is employed to warm air taken directly from the outer air. The air is warmed by the iron back of the stove and smoke-flue, upon both of which several broad flanges are cast, so as to obtain a large surface of metal to give off the heat. This giving-off surface (amounting in the case of No. 1 grate to about 18 square feet) is sufficient to prevent the fire from ever rendering the back of the grate so hot as to injure the air which it is employed to heat. The fresh air, after it has been warmed, is passed into the room near the ceiling by the flue shown in the woodcut.

The flue which has been adopted for barracks is carried up by the side of the smoke-flue in the chimney-breast.

In order to afford facilities for the occasional cleansing of the air-chamber, and those parts of the air-channels connected with it, the front of the stove is secured by screws, so that it can be easily removed, thus rendering the air-chambers accessible.

The stove was designed with the object of being applied to existing chimney-openings. In so applying it, the air-chamber is to be left as large as possible, thoroughly cleansed

K

from old soot, rendered with cement, and lime-whited. Should the fireplace be deeper than 1 foot 6 inches, which is the depth required for the curved iron smoke-flue, then a lining of brick-work is to be built up at the back to reduce it to that dimension. The chimney-bars, if too high, must be lowered to suit the height of the stove, or to a height above the hearth of 3 feet 3 inches; they must also be straightened, to receive the covering of the air-chambers. These coverings should be of 3-inch York or other flagging, cut out to receive the curved iron smoke-flue, and also to form the bottom of the warm-air flue in the chimney-breast; or the covering may be formed of a brick arch. In new buildings the air-chambers may be rectangular; they must be 4 inches narrower than the ex-treme dimensions of the moulded frame of the stove, so as to give a margin of 2 inches in width all round for a bedding of hair mortar. In existing buildings, the recess in which an ordinary firegrate would be fixed forms the chamber in which the air is warmed. Great care must be taken in bedding the several joints to prevent smoke from the flue passing into the fresh-air chamber or fresh-air flue.

The mode of admitting external air into this chamber must depend upon the locality. If the fireplace be built in an external wall, the opening for fresh air can be made in the back; but if an internal wall, it will be necessary to con-struct a channel from the outside, either between the flooring of the room and the ceiling-joists of the room below (if there be independent ceiling-joists), or between the floor-boards and the plaster ceiling of the room below, in the spaces between the joists; or by a tube or hollow beam carried either below the ceiling of the room altogether; or, as is often more convenient, behind the skirting of the room in which the firegrate is placed.

In this country these horizontal ducts should contain a sectional area, for the large size grates of 84 square inches, for the medium size of 60 square inches, for the smaller size of 36 square inches ; the clear area through the grating covering the opening to the outer air should be equal in area with that of the flue.

If the flues are of considerable length, and with bends, the sectional area should be rather more than that mentioned, to allow for friction ; but if there be a direct communication with the outer air, the sectional area may be rather less than that recommended. In exceptionally cold weather it may be advisable to reduce temporarily the area of inlet.

The amount of air delivered through the fresh-air flue varies somewhat with the direction of the wind. The inlet shaft acts best when the windows, doors, and other inlets to the room are closed, as it then becomes the sole inlet for fresh air for the room.

In the ventilation of barrack-rooms or hospitals, it was not intended that the fresh air warmed by the grate should be the whole supply of fresh air, nor that the chimney should be the sole means employed for the removal of the air to be extracted. In ordinary houses, however, the grate, if adopted. might be used in such a manner as to perform the whole functions of ventilation. In this case it is of course necessary to remember that the ventilating power is a fixed quantity, and that in originally settling the size of a grate for a particular room it will be necessary to bear in mind the general object for which the room is to be employed, and the number of occupants for whom it is required to furnish efficient ventilation. Experiments show that no room can be considered even tolerably ventilated as a permanent arrangement unless at least 1000 cubic feet of air per occupant are renewed per hour ; consequently a room 20 feet long by 15 feet wide and 10 feet high (i. e. with 3000 cubic feet of space), with three people in it, would not require the air to be changed much more than once an hour ; whilst, if occupied by twelve or fourteen people, it would require to be changed five times in an hour. If the normal use of the room was for three people it would not be worth while to provide for the extra number by which it might on occasions be occupied, as their wants in such a temporary case could be met by opening the windows slightly at the top.

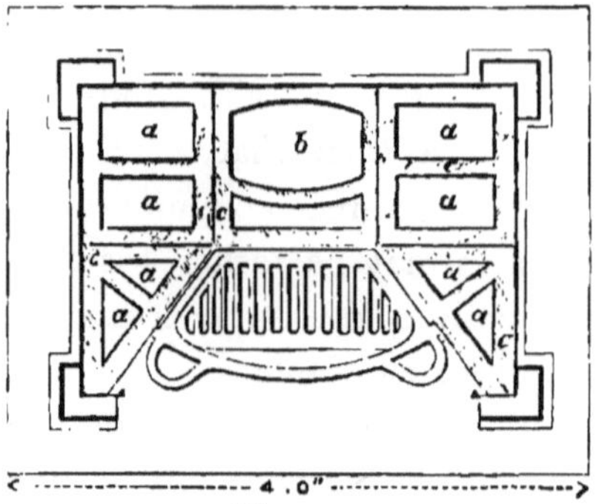

Fig. 25 a.

There is one point connected with the fresh-air flue which must be carefully attended to—viz. the fresh air should be taken from places where damp or impurities cannot affect it, and the flue must be so arranged and constructed as to afford easy means of being periodically thoroughly examined and cleaned. In barracks the rule is that such cleansing should take place at least once a year.

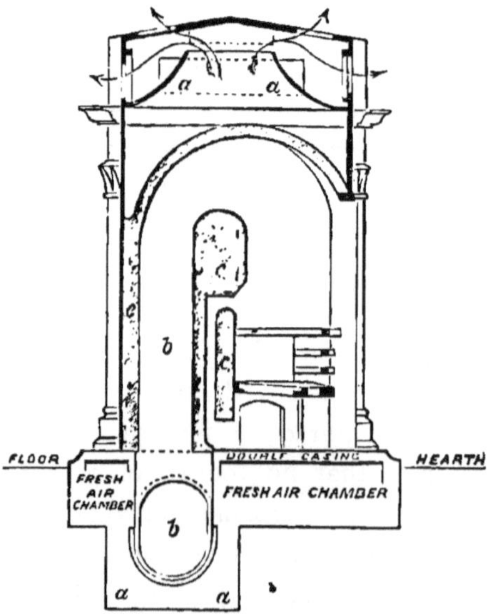

a a. Fresh-air Flues.   *b b.* Smoke Flue
c c. Fire Clay.

Fig. 25 b.

The area of the grate of No. 1 stove (Fig. 25a) is 84 square inches, of which 58 are solid, and 26 afford space in the centre for the passing of air. The front is open, and air is passed on to the coal from the back in the manner already described. The grate will contain about 18 to 20 lbs. of coal; when the fire is maintained for from twelve to fifteen hours, a total consumption of about 2·5 lbs. per hour, or 40 lbs. for sixteen hours, will suffice to maintain a good fire.

In new buildings it would be possible, and in some cases it might be desirable, to extend this heating surface considerably by carrying up the smoke flue inside the warm-air flue. But care must be taken not to overheat the air. This plan has been adopted in the fireplaces for the wards of the Herbert Hospital, where the fireplace is in the centre of the ward, and the chimney consequently passes under the floor, and is placed in the centre of the flue which brings in the fresh air to be warmed by the fireplace; by this means more than 36 superficial feet of heating surface have been obtained for warming the fresh air in addition to the heating surface afforded by the air flues in the fireplace. (Figs. 25*a*, 25*b*.)

The fire stands in an iron cradle fitted to the fire-clay back and sides, and a current of air passes. from the ward under the grate through the slit in the fire-clay at the back, where it becomes heated, on to the top of the fire to assist the combustion, and thus prevent smoke. The top of the stove is coved inside, to lead the smoke easily into the chimney. The main body of the stove is a mass of fire-clay, with flues cast in it, up which the fresh air passes from the horizontal air flue already mentioned, in which the chimney flue is laid. Thus all parts of the stove employed to warm the fresh air with which the fire has direct contact, are of fire-clay. This is especially essential in hospitals, where every element of possible impurity of air should be avoided. The sectional area of the fresh-air flue with this arrangement of grate may be 1 square inch for every 100 feet of cubic contents of the spaces to be warmed, for favourable situations; but in cold or exposed localities a less area may be allowed.

The horizontal chimney flue in the Herbert Hospital fireplaces is formed of two layers of sheet iron, separated by a thin layer of fire-clay, so as to prevent overheating of the surface, and it is about 110 square inches in area. The horizontal chimney flue terminates in a vertical flue in the side wall, which should be rather larger in area than the horizontal flue. This vertical flue is carried in the upper floors to a height of double the length of the horizontal flue, and is carried down to the basement, whence it can be swept. The points of connection between the horizontal chimney with the descending flue from the fireplace, and with the ascending flue in the wall, are very carefully rounded, as this is essential to

A. Fire lump with warm-air flue through back.
B. Warm-air pipe to fit into socket on hob, in lengths of 1 ft. 3 in. each.
C. Bend to fit socket of the above pipe.
D. Mouthpiece with louvre front to fit on bend. One of these, 6 in. long, is supplied with each range.
Increased heating surface for warming the fresh air is provided by means of a grating inside the socket at E.

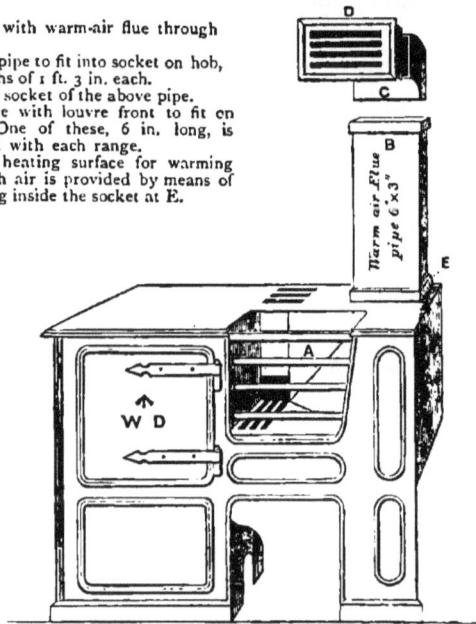

Fig. 26.

assist the passage of the smoke. The horizontal flue is swept from an opening, to which access is obtained by taking up a moveable board in the floor, and by pushing a brush along the flue, and thus forcing the soot into the vertical flue, whence it falls down and is removed at the opening in the basement.

There is placed a spare flue by the side of the vertical flue, terminating in a fireplace in the basement, which enables the vertical flue to be warmed, so as either to make it draw when

the fire is first lighted, or to enable a current to be maintained for ventilating purposes through the fireplace when the
fire is not lighted. The portion of floor over the horizontal
flue should be so constructed as to be taken up in order to
enable the air flue to be easily and thoroughly cleaned
periodically.

The principle of these arrangements for utilising to some
extent the heat in the chimney has been adopted for barracks
in the case of grates for married soldiers; these would be
useful as cottage grates. They have a small oven, and an
open fire; warmed air is introduced into the room by means
of an iron flue carried up from the fire-brick lining of the stove
inside the chimney, and introduced into the room near the
ceiling through a louvred opening; by this means the heat of
the smoke is utilised to warm fresh air admitted to the room,
and prevent draughts.

This description of grate was devised for the purpose of
combining a sufficient power of cooking for a cottage with
great compulsory economy of fuel (see Fig. 26).

It must however be observed, that in proportion as heat
is absorbed from the fire for warming inflowing air, and
the combustion of the fuel checked, so is the draught, i. e. the
effect of the chimney as a pumping-engine to remove the air,
diminished.

The limit to which the heat from the fire can be utilised
will be the point at which a sufficient amount of heat is left
in the chimney, to cause an adequate draught, so as to ensure
the combustion of the fuel.

With these ventilating fireplaces in temperate weather,
when windows or other inlets are opened direct from the
fresh air, the entrance of warmed air will be checked, whilst
the ventilation of the room would be continued; and if desired,
the inlets for warmed air may be provided with valves to be
closed when the fireplace is wanted rather for ventilation than
for warming.

Numerous experiments have been made on the ventilating fireplace for barracks. These experiments show that the air is generally admitted into the rooms at a temperature of about 20°, or from that to 30° Fahr. above that of the outer air. The design of the grate was intended to preclude the possibility of such a temperature as would in any way injure the air introduced; and the experiments made by the late Dr. Parkes in a hospital ward at Chatham, in April 1864, illustrate the hygrometric effect with the grate in use. The difference between the dry and wet bulbs in the ward varied from 8° to 5°, viz. on the 17th of April, 8.5°; on the 18th, 6°; on the 19th, 5.5°; on the 20th, 6.5°; on the 21st, 5.0°. On examining the record of the dry and wet bulbs, no evidence was seen at any time of any unusual or improper dryness of the atmosphere. The difference between the two bulbs was certainly always greater in the ward than in the outer air, but the difference was not material. The temperature of the rooms was invariably found to be so equable, that when the grate was in full action, and the windows and other means of ventilation closed, thermometers placed in different parts of the room, near the ceiling and floor, in corners furthest from the fire, and on the side nearest to it, but sheltered from the radiating effect of the fire, did not vary more than about 1° Fahr. The variation of temperature in different parts of a room warmed by an ordinary fire, by radiation, without the action of warmed air, will be found to be from 4° to 6° Fahr., and sometimes even much more in cold weather.

General Morin, with the object of utilising the grate as the sole means of ventilation for a room, lays down the principle that the whole of the air shall be renewed five times in the hour. To perform this effectually, it is necessary that the area of the chimney outlet shall afford about one square inch of area for every 100 cubic feet of content of the room, and that the area of the fresh air inlet at its opening into the room should be enlarged to afford about 14 square inches for every 100 cubic feet of content of the room ; so as to prevent the air entering with a rapid current. But on an average this quantity of air is more than is necessary. The Barrack and Hospital Improvement Committee's proposal would resolve itself into this—viz. that the air in barrack-rooms should be completely changed about twice in an hour (inasmuch as they required a cubic space of 600 cubic feet per man), and for all ordinary purposes this would probably suffice ; as, however, this proposal was based on a limited number of occupants, with a more crowded room the amount must be increased.

General Morin's experiments in 1864–5–6, with fireplaces constructed in the ordinary manner, and with others on the plan above described, in which the chimney was utilised for warming the air, showed that whilst with an ordinary fireplace the heat which is utilised in a room amounts only to one-eighth of the heat given off by the coal, or ·125, in these fireplaces the heat utilised in the room was ·355 of the heat given off by the coal, or one-third ; therefore, to produce the same degree of warmth in a room, this grate requires little more than one-third of the quantity of coal required by an ordinary grate. The ventilation of the rooms was at the same time effected by passing a volume of air through the room in one hour equal to five times the cubic contents of the room. An equable temperature was maintained during the experiment. There was no perceptible draught, and although the doors fitted badly, scarcely any air was drawn in through the crevices.

In conclusion, the merits which are claimed for this ventilating fireplace are :—

1. That it ventilates the room.

2. That it maintains an equable temperature in all parts of the room, and prevents draughts.

3. That the heat from radiation is thrown into the room better than from other grates.

4. That the fire-brick lining prevents the fire from going

out, even when left untouched for a long time, and prevents the rapid changes of temperature which occur in rooms in cold weather from that cause.

5. That it economises fuel partly by making use of the spare heat, which otherwise would all pass up the chimney, and partly by ensuring by its construction a more complete combustion, and thereby diminishing smoke.

6. That it prevents smoky chimneys, by the ample supply of warmed air to the room, and by the draught created in the neck of the chimney.

CHAPTER XII.

THE open fireplace is convenient in ordinary living-rooms in this climate, because the weather is generally so temperate that it is only exceptionally necessary to provide much artificial warmth. In climates with colder winters the open fireplace becomes expensive, and is used rather as a luxury, because every room and corridor requires to be warmed, and if this is done by open fireplaces the carrying of the fuel alone in such cases becomes a serious inconvenience.

In order to ensure an adequate change of air in cold climates when there is no open fireplace, and when the weather requires closed windows, and in hot climates when the movement of air is very small, the fresh air may be introduced by air-flues connected with fans or pumps, or its inflow may be dependent, as it is in the case of the open fire, upon the action of heated extraction-shafts.

This, however, has not hitherto always been the practice. In many cases, rooms are warmed by close stoves on the French or German plan, or by hot-water pipes, without any arrangements for introducing fresh air. On the other hand, when the warming is effected by an inflow of warm air, the air is generally brought in by special channels from a heating apparatus placed in a central part of the building. This latter plan is much used in the United States, and in Canada, where the winters are very cold. The warmed air is supplied (generally at a high temperature) from a heating apparatus which is usually placed in the basement or lower part of the house.

The effect of introducing warm air at the lower part of a ward on each side, and allowing it to escape at the top, was illustrated by a series of careful experiments, by Professor E. Wood of Harvard University, upon the various currents prevailing at the same time in a hospital ward at Boston, U. S. (Fig. 27.)

The ward was 96 feet long by 26 feet 3 inches wide, with seven opposite windows, the tops of which were 16 feet above the floor-level, and 14 beds on a side. The ceiling was arched, 20 feet high in the centre, and averaged a height of about 18 feet, without any obstructing cornices. The floor-space of the beds was 88.65 square feet, and the cubic space per bed 1629 feet. The fresh air was admitted warmed at a temperature of 90° by openings 1 foot square each, under each window, just above the floor-level, 14 openings in all, and the vitiated air was extracted by five openings, each 3 × 6 feet in the length of the ward in the centre of the ceiling. The hourly supply per bed was 9000 cubic feet. The high temperature of the inflowing air caused it to enter with velocity. This velocity was soon lost. There was an upward movement of air throughout the ward, but a comparative stagnation in the centre, for a height of about 8 feet from the floor-level—and also a comparative stagnation commencing over the bed-heads and extending to the upper part of the ward, as shown in the diagram. The respiratory impurity, as shown by excess of CO_2 over outer air, at the lower level was ·00131, and at the upper part ·00240—thus showing that with this construction of ward and system of ventilation, there was a want of adequate circulation of air, as well as a disadvantage in the height above the top of the windows, which were 14 feet in height; openings at the sides higher up would have permitted the foul air to escape more rapidly from the room, instead of occupying the waste space above that level at the risk of cooling and falling again and remixing with the air of the room. This example shows that extraction-shafts should be so placed as to cause a circulation of air such as is affected

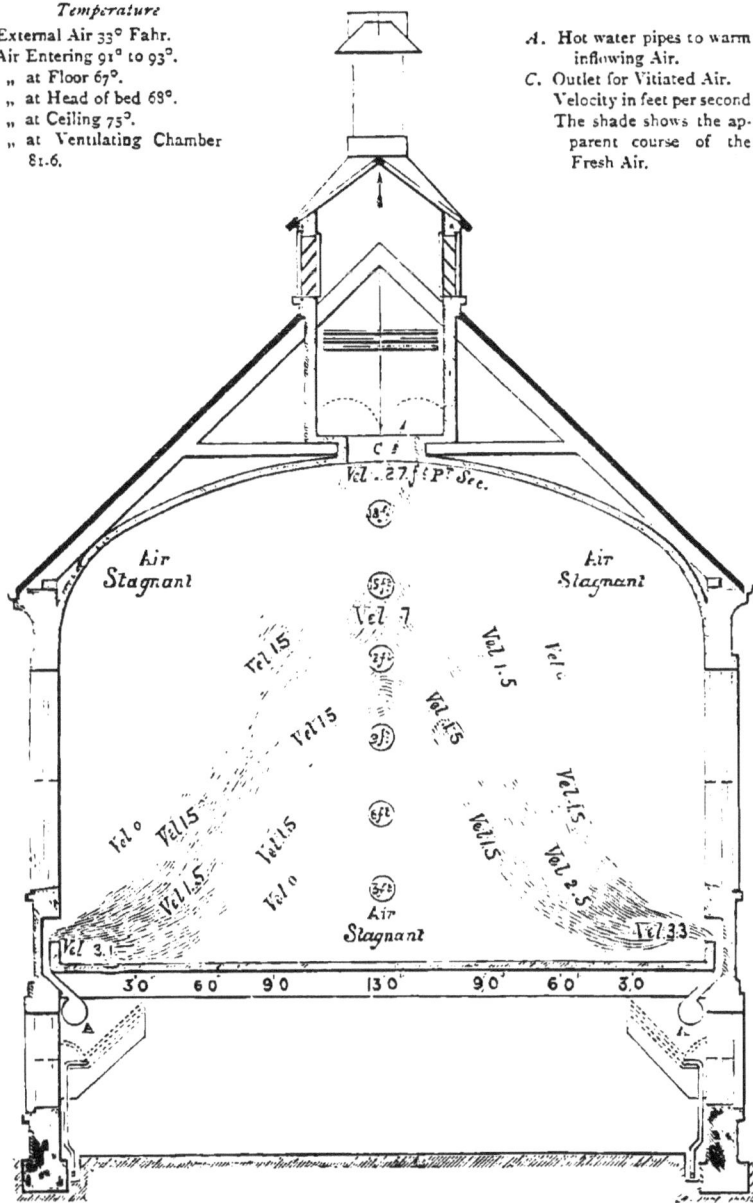

Temperature
External Air 33° Fahr.
Air Entering 91° to 93°.
„ at Floor 67°.
„ at Head of bed 69°.
„ at Ceiling 75°.
„ at Ventilating Chamber
81·6.

A. Hot water pipes to warm
inflowing Air.
C. Outlet for Vitiated Air.
Velocity in feet per second
The shade shows the apparent course of the
Fresh Air.

Fig. 27. Experiment in a hospital ward at Boston, U.S., by Professor E. Wood of Harvard University.

by a fireplace, which acts to cause a circulation, and to extract the air.   In England the system is often resorted to in asylums and large houses, of warming the air, and supplying the warmed air by flues to different parts of the building ; this should be combined with some plan which, whilst causing circulation, will also extract the vitiated air.

The method of warming rooms by close stoves or by hot-water pipes in the room, do not of themselves necessarily entail any change of air.

In Germany and the northern parts of Europe, where close stoves are used, the stove is generally filled with fuel once a day at an opening often outside the room, and no removal of air takes place by means of the stove.   Therefore, where there are no special means for changing the air, the rooms become close and unhealthy ; but in very cold weather, when there is a great difference between the temperature indoors and out, a considerable change of air is effected through crevices of doors and windows, and through every available aperture ; moreover a spontaneous change of air will take place through the walls, when the inside temperature greatly exceeds that out of doors.

Dr. Böhm (than whom no one has better studied ventilation) adopted many years ago the following system in the Rudolf Hospital at Vienna, where the wards are heated by close stoves.   He there warms fresh air by means of passages constructed in the fire-clay stoves, placed within the ward, and the fresh warm air passes into the ward from the top of the stove.   He provides flues of a large size, and proportioned to the size of the ward, carried from the level of the ward floor to above the roof, inside which the flue from the stove is carried, and in cold weather the difference of temperature between the air in the flue and the outer air causes a sufficient current in these flues to ventilate adequately the ward. By this means the fresh warmed air, instead of passing off to the upper part of the ward, and thence away by flues, is made to circulate towards the floor of the ward, thus bringing into action one of the conditions of the open fireplace; each ward is self-contained in respect of its warming and ventilation, as is the case when an open fireplace is used. This is for winter ventilation.

In warm weather, when the stove is not in use, the outlet near the floor is closed, and an outlet into the extraction-shaft near the ceiling is opened by means of a valve, so as to allow of the escape of the air from the upper part of the room. The flues in the stove remain as inlets, and other inlets for summer use direct from the open air are provided at the floor-level and elsewhere; and the windows can be opened if desired.

There have been numerous close stoves invented of late years to supply fresh warmed air; in France the system of hot air stoves prevails extensively, and it is being gradually adopted in this country. These provide for the admission of fresh air : but care is not always taken to provide both for the removal of the vitiated air and the supply of fresh air; and without provision for both good ventilation is impossible.

In this climate sufficient ventilation can generally be obtained in private houses, in hospital wards, in barrack-rooms, and in workhouses, when the proportion of floor-space allotted to each occupant is comparatively large, by providing extraction flues and inlets as already described for the purpose of taking advantage of the difference between the temperature indoors and the temperature outside, and by a judicious use of open windows. But in prisons, asylums, and buildings occupied by persons under strict rule, as well as in buildings where large numbers are congregated together for a limited time, such as theatres, churches, legislative assemblies, public meetings, crowded dining-rooms or smoking-rooms, and otherwise, adequate ventilation can as a rule only be secured by the adoption of some means for extracting the air, beyond that which would be provided by the simple form of outlet and inlet above described.

This may arise from various causes. For instance, in densely occupied buildings, where the volume of air to be removed is large, and the building complicated in shape, dependence on the difference of temperature and the action of the atmosphere alone might require inlets and outlets of an inconvenient size, and therefore some power of extraction to accelerate the current is necessary. There are also many other cases in which it is necessary to resort to mechanical means for changing the air. In all such cases, ventilation must be combined with warming, on the score of convenience as well as of economy.

In hot climates the conditions differ materially from those already alluded to. The difference of temperature indoors and out is rarely sufficient to produce an adequate change of air. The object to be attained is the reverse of that sought in this climate. The fresh air to be supplied should be cooled down below its outside temperature instead of being raised above it.

In hot climates it will therefore generally be necessary to provide some artificial means of removing the air, in order to secure an adequate change of air in rooms or halls occupied by many people.

Ventilation by mechanical means may be effected either by propulsion or by extraction. By propulsion fresh air would be forced into the building to drive out the vitiated air. To effect this, Dr. Arnott proposed an apparatus on the principle of the gas-holder, by which he would carefully regulate the quantity of inflowing air. In the House of Commons there is an apparatus on the principle of a large pump, for occasional use, by which in the summer a given quantity of cool air can be forced into the House. At the Hospital Necker at Paris, fans were used for forcing a given quantity of air into the wards ; and propulsion has been adopted in some recently constructed English hospitals.

It is probable that a system of propulsion might be found to be the more convenient system for freshening the air in the rooms in hot climates, combined with openings in the upper part of the room. The air might be compressed in the course of the propulsion in impervious underground channels, in which it would be retained sufficiently long to be cooled down by the lower underground temperature, and in passing into the rooms or wards it would be still further cooled down by the expansion. In towns where the subsoil is liable to be very polluted greater care must be exercised to prevent the ground air from mixing with the air for ventilation.

But, as has been already observed, experience shows that a system of propulsion has not acted satisfactorily in hospitals in this climate, unless it has been combined with a system for the extraction of the vitiated air. And under ordinary circumstances, where extraction is resorted to, an adequate quantity of fresh air can be drawn in by the operation of the extraction-shafts, without the additional expense and trouble of propulsion.

In the case of ventilation by extraction, it is generally found convenient and economical to provide one principal extraction-shaft, to which the extraction-flues from the separate apartments are brought; and to warm the fresh air at some central apparatus before it is drawn into the rooms. The effect to be sought for in a room from the use of extraction-shafts is as follows :—

Currents created by outlets placed near the floor-level and led into extraction-shafts must be arranged to move only at a low velocity into the outlet, in order to prevent draughts : the temperature of air as it enters the outlet is not affected, as is the case with the open fireplaces ; therefore the tendency to create a circulation of air in the room does not prevail to the same extent as with an open fireplace ; and if a room without an open fireplace is to be equally well ventilated by extraction-shafts alone, the advantages which an open fire presents must be compensated for by placing the outlets and inlets in such positions and in such numbers as will prevent stagnation of air in any part of the room. In cold weather the extraction should generally take place from the lower part of the room, so as to draw off the cold air from near the floor, and cause the warm air above to circulate.

In this case the orifices for extracting the air should be raised above the floor sufficiently to prevent dirt falling into them, and it is advisable if possible not to place them near the feet of persons sitting down in the room, because the continual flow of air, even at a low velocity, produces an evaporation which in time may create a feeling of cold. It may be generally said that the larger the area through which the air is drawn off the less will any unpleasant effects be experienced.

Thus in a lecture-room the vertical risers of the seats may be advantageously made entirely of open grating, so that the whole area is available for drawing off the air.

On the other hand, in rooms with many occupants, where much gas is used, in hospital wards, in smoking-rooms and dining-rooms, or other places under similar conditions of the vitiation of air, the heated air at the top of the room should be rapidly removed by special outlets near or in the ceiling, in addition to extraction at the floor-level; care being taken that both sets of outlets shall draw efficiently. The heat which results from gas affords of itself a means of extraction, if judiciously applied.

Inlets for the supply of fresh air should be placed above the heads of the occupants of the room, and admitted through inlets sufficiently far from the extracting outlets to prevent its being carried away at once.

The construction of the large cathedrals lends itself to ventilation; because the lofty nave with clerestory windows, and the towers, are convenient means for the removal and admission of air, and if the windows were all opened after a service, very little inconvenience would be experienced even from the largest congregation. But there are many churches in this country where large congregations are crowded into buildings of inadequate size and height, and these require special care.

In churches with galleries, densely occupied, and in halls for meetings, the air should be drawn off from the top as well as from below. The upper air, in cases of crowded meetings, is filled with much moisture from the breath ; it is therefore advantageous to warm the air thus drawn off as it enters the extraction-shafts, in order to increase its capacity for moisture, so that the moisture shall not be deposited before it passes into the outer air.

Theatres present peculiar difficulties of ventilation, from their complex arrangement. The entrance, staircases, and corridors should be warmed independently of the part occupied by spectators. These require but little ventilation. The part occupied by spectators is under different conditions when the curtain is down and when it is raised ; therefore the stage should be warmed and ventilated independently of the part reserved for spectators. As the latter part requires a large amount of light, and as light is inseparable from heat, the light may be utilised for promoting ventilation. The heated air should be drawn off, and the fresh air, duly warmed in winter, should be admitted, at levels, and in a manner not to inconvenience spectators with draughts. Adequate ventilation in a theatre will, as a rule, be more satisfactory, if some mechanical appliance be provided to force in the fresh air as well as to remove the vitiated air. In summer the ventilation requires to be modified so as to supply air cooled either by compression or otherwise, as may be found most advantageous. The effectual ventilation of a theatre requires the presence of a special attendant to watch continuously the condition of the air in the theatre, and modify the arrangements from time to time.

The object of this treatise is not to define special methods of ventilation ; but rather to bring into view the principles which should guide the architect or engineer in considering

the system most applicable to the requirements of his particular case; because the enunciation of specific rules in so
complex a subject might fetter him and prevent advancement
in sanitary science.

The following instances of some of the systems of ventilation which have been adopted in various buildings will afford
a general idea of the principal points which have to be considered.

The old Roman plan, by which the source of warmth was
uniformly distributed by heating the floor and walls of a
room, gave an equable temperature; and by warming the
floor and walls rather than the air partially provided one of
the advantages arising from the open fire.

If in connection with this system fresh cold air were admitted near the ceiling, and if the warmed vitiated air were
removed by conveniently placed outlets, and if each occupant
had a sufficient floor-space, it is probable that this would form
a very comfortable arrangement.

The Roman floor was a floor of tiles laid hollow with flues
underneath. The furnace, with a fire of wood or charcoal,
was formed at one end, and the smoke was carried backwards
and forwards along the flues under the floor, and finally either
led up the walls or to a chimney. This chimney was generally
so placed as to act as a means of keeping the store of wood dry.

But such a plan could only be adopted in a specially
constructed building.

An experiment was made to compare the economical effect
of warming by means of a heated floor with the effect of heating by means of a ventilating fireplace; the experiment lasted,
with each mode of warming. for two days. It showed that, in
the case of the warmed floor, the room was maintained at a
temperature of about 18 degrees above the temperature of the
outer air with an expenditure of 56 lbs. of coal and 112 lbs. of
coke, whilst with the ventilating fireplace the expenditure was
only 75 lbs. of coal; the cost being 3s. 4d. for the warmed floor

as compared with 1s. 4d. for the ventilating fireplace. It is probable however that in a house designed and constructed specially for this mode of heating, better results would have been obtained.

The ventilation of Derby Infirmary, designed and carried out more than sixty years ago, is an instance of ventilation by natural extraction and propulsion alone.

Flues were led from each ward to a central shaft, at the top of which was placed a cap with a vane arranged so as always to cause the opening to face away from the wind. Fresh air was introduced by means of another shaft, also furnished with a cap with a vane arranged to make the opening always turn towards the wind like a wind-sail in a ship. The fresh air entering this latter shaft was carried down underground along a channel 70 feet long to an iron stove covered with flanges, where it was heated in cold weather ; and thence it passed up by separate flues to the wards. The extraction-flues were led into the central shaft from apertures near the ceiling, and the wards were also provided with open fireplaces near the floor-level. Thus the action of the wind was employed to extract the vitiated and to propel the fresh air. The amount of air which was thus passed through the wards with a velocity of wind of about 3 miles per hour, was recorded as 400,0c0 cubic feet per hour. But of course on perfectly calm days the effect of the extraction-shaft was limited to the extracting force resulting from the difference of temperature and the height of the shaft ; this force was necessarily much modified by the friction due to the length of the flues in addition to the delays caused by the horizontal portions of the flues. Straight flues carried up separately from each ward would have been more efficient for extraction. The underground channel had the effect of partially warming the fresh air in winter and cooling it in summer [1].

---

[1] This system was invented by G. W. Sylvester and William Strutt, but recent alterations have abolished it.

The Great Opera at Vienna is ventilated by extraction-shafts carried from round the ceiling, from above the central chandelier and from the upper part of the boxes and galleries. These are brought into action by the heat generated by hot-water boilers and by the products of the gas chandelier; the admission of air is regulated by propulsion by a fan from below. The air thus supplied is partly passed over pipes heated by hot water in winter, and is partly left cold. Channels for hot air and channels for cold air are provided to all parts of the theatre, under each occupant of the pit and stalls, and in the corridors for the supply of the boxes in addition to a large volume of air admitted near the stage so as to keep the ventilation for the spectators separate from that for the actors. In hot weather the air can be cooled by being passed through spray. In order to maintain the ventilation in efficient action, metal thermo-electric meters are placed in different parts of the theatre, all connected by electric wires with a small office in the basement, and these indicate the temperature in different parts of the building at each moment on a board in the office. When the observer sees that the temperature is raised or falls unduly, in any part of the house, he can open or close valves, which are under his hand, and which regulate the quantity of warm or cool air admitted to each part of the house, and thus alter the temperature as desired.

The New Opera at Paris is warmed by hot-water apparatus for all the parts behind the stage, on the assumption that this method does not dry the air so much, whilst the part of the theatre occupied by the public is warmed by stoves, as being more rapidly put in action.

The fresh air is introduced through flues in the floor of the several tiers of boxes, and the vitiated air is extracted partly from the back of the boxes and partly from the floor of the pit and stalls. The heat from the lights is not applied to ventilation.

Figure 28, on the next page, shows the system in use in the Houses of Parliament.

The Houses of Parliament are warmed and ventilated by a combined arrangement. For the House of Lords and the House of Commons the arrangements are identical.

The fresh air is admitted into a lower chamber, where, after being filtered through screens of jute and cotton wool, it is warmed by pipes. Through these pipes steam circulates; broad vertical flanges are cast at intervals on the pipes in order to increase the heating surface. Each of these flanged portions of pipes has therefore a similar heating power. The frequent alteration in the number of occupants of the House of Commons makes it necessary frequently to alter the temperature of the air. This alteration is effected by an assistant, who watches the thermometer and covers with pieces of woollen fabric, or uncovers, a greater or less number of these flanged portions of the pipes, so as to diminish or increase the heating surface as may be required.

The fresh air is supplied to this chamber from the adjacent courtyards, which are covered with asphalte and kept tolerably clean.

The air, after having been warmed in the lower chamber, is passed through four large circular openings of about three feet six inches diameter each, into a chamber above : this chamber is practically a portion of the House of Commons, and is separated from it only by means of an iron grated floor covered with open matting. The grated floor and the grated risers of the steps on which the seats are placed are used as inlets. Hence much of the care taken to purify the air in the lower chamber is rendered nugatory by passing the cleaned air through the floor mats, which are being constantly fouled by the dirt brought in on the boots of members. The glass ceiling of the House of Commons has openings into the space in the roof, from which a channel leads down under the basement to the foot of the clock-tower, where a large fire

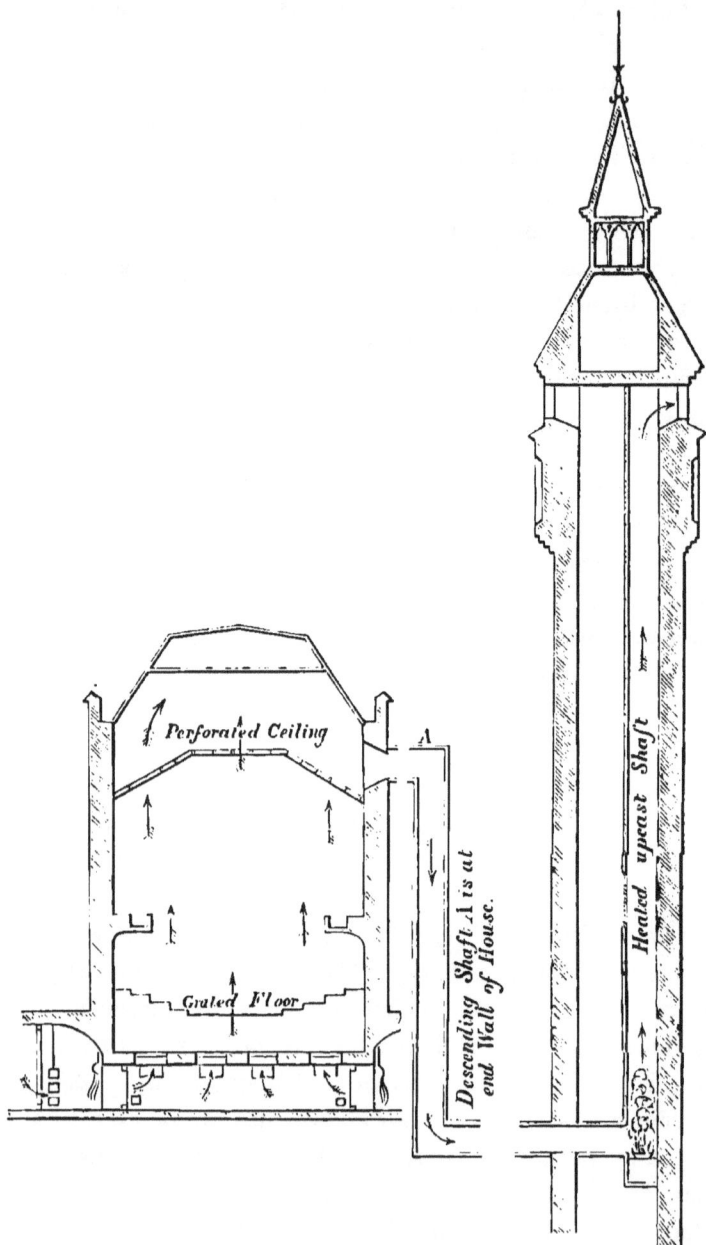

Fig. 28. Ventilation of the House of Commons.

is maintained ; and this forms the exhaust by which the air is drawn through the house; as much as 1,500,000 cubic feet have been passed through per hour. If the house were full, this would represent somewhat under 2000 cubic feet per occupant per hour, including members, attendants, and strangers. The lighting is effected by gas-lights placed above the glazed ceiling.

In summer an arrangement has been adopted, by which air, made cool by passing over ice, has been forced into the house. This is an expensive process for cooling the air. Equally good results at a much less cost might be obtained by utilising the underground temperature in summer, or by using compressed air. The air is drawn from the ground-level, which is an undesirable arrangement. Much purer air could be obtained by bringing the air down from the lofty towers which form part of the structure.

The French Legislative Chambers were ventilated by General Morin on a plan the reverse of that adopted in the Houses of Parliament. The Legislative Chamber has a cubic content of about 400,000 cubic feet ; the arrangements are intended to provide for 1,060,000 cubic feet of air to pass through the chamber in an hour. The air is extracted through openings at the floor-level, and others placed in the vertical risers under the seats, in the body of the hall and in the galleries ; there are also outlets in the staircases and corridors of approach. The united area of the extracting outlets is nearly 160 square feet, which would allow of a velocity of 1·8 feet per second in supplying the regulated quantity of air.

The air is admitted at the ceiling-level through openings in the cornice all round the chamber and in the capitals of the columns. The total area of inlet is about 195 square feet, which would allow of a velocity of about 1·5 feet per second for the inflowing air, with the maximum supply, which however is rarely attained.

The fresh air is brought down by a flue from a height of 60 metres into a distributing or heating chamber at its base, where it is warmed to the necessary temperature by four (or fewer) stoves of brick provided with air-passages. The adjacent committee-rooms, and the rooms and corridors for the use

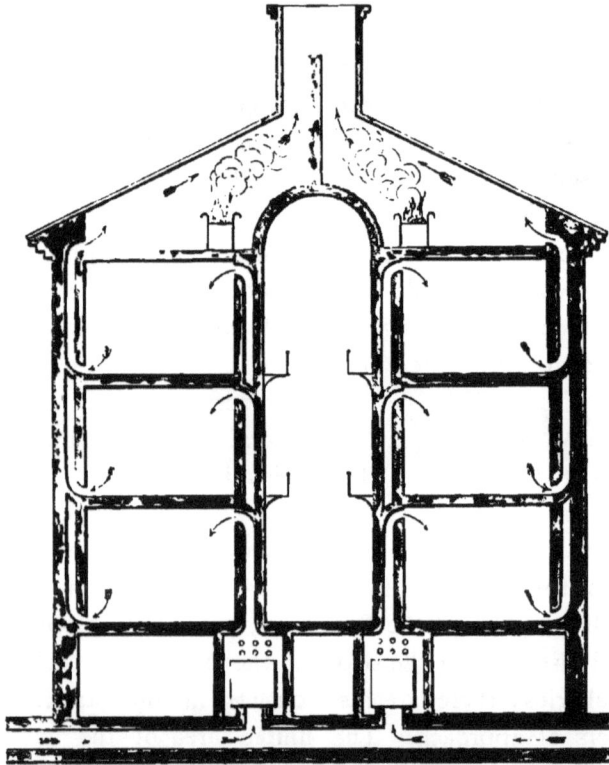

Fig. 29.   Sir Joshua Jebb's system of Ventilation for Prisons.

of members, are warmed and ventilated on the same principle as the Legislative Chamber—the total amount of air being 1,600,000 cubic feet per hour. The difficulty of maintaining a uniform temperature in the Legislative Chamber arises from the continual variation in the number of occupants. The temperature maintained in the winter is about 64½° Fahrenheit,

with a variation seldom exceeding 3° or 4°. In spring, with an
outside temperature of 57°, the inside temperature has rarely
varied more than from 64° to 68°. In summer, with an external
temperature of 82°, a temperature has been maintained in the
chamber of from 73° to 77°.

The air being brought first into the vaults below the ground,
its temperature is raised to begin with by the temperature
of the vaults ; in winter as much as from 9° to 12° above the
outside air ; in summer it could be cooled down as much
as from 16° to 18° below the outer air, but this would be
beyond what would be endurable in practice.

The temperature in the main extraction-shaft averages from
36° to 45° above that of the outer air ; but in summer, with
an out-door temperature of 83°, it was found necessary to
raise that of the main extraction-shaft to about 142°, or a
difference of 59°. Experiments were made to ascertain
whether, in order to cool down the chamber in the great
summer heats, it was desirable to continue the ventilation
through the night, but the results did not correspond with the
extra expense.

The system of ventilation of cells in prisons adopted by
Sir Joshua Jebb was practically the same as the system
originally proposed by Sylvester and General Morin, viz. the
extraction of the air at the lower part of the cell and its
admission near the ceiling (Fig. 29). In cold weather this
method of admitting the warmed air keeps the temperature
of the cell uniform. When the extraction-current is sluggish,
as may be the case in warm weather unless much extra fire is
used, this plan of removing the air from below does not always
sufficiently relieve the cell unless the window is opened. A
better effect will be produced by the removal of air from the
upper part of the cell. The most convenient arrangement
would be to have a shaft as shown in the diagram (Fig. 30),
with an opening to be closed alternately either at the top
or the bottom according to the weather. In large prisons the

most economical arrangement is to warm the fresh air at some central source of heat and distribute it thence over the building. The extraction must either be mechanical, or by means of a heated shaft, because these prisons have large central halls into which the cells all open, and which would disturb the ventilation in the absence of mechanical extraction. Care should be taken that the flue from each cell shall have a sufficient length before it enters the main extraction-shaft, to ensure that there are no reverse currents, for these might bring impure air back into the cells. The arrangement requires that the fire for the extraction be kept up day and night. For cells on one floor, such as barrack or police cells, where the number is small, the simplest and most efficient arrangement is to provide a small shaft for each cell direct to the open air, with an opening near the floor for winter and near the ceiling for summer, one valve being arranged to be closed when the other is open, the cells in this case being warmed by hot-water pipes in coils, or flanged pipes, behind which fresh air is admitted into the cell so as to come in warmed when the pipes are in use. Direct openings from such cells into a general extraction-shaft carried along the ceiling of the several cells are objectionable, because under such an arrangement the impure air from one cell frequently passes into the adjacent cells. If such a general extraction-shaft be provided, the least dangerous plan would be to place it at a height of at least 6 feet above the ceiling of the cells, and to carry a small shaft from each cell vertically up for a few feet in length into it ; a current being maintained by mechanical means, or by heat in the shaft.

Ceiling.

Valve flap, closed at top when bottom is open, and *vice versa*.

Flue

Valve flap, open at bottom when top is shut, and *vice versa*.

Floor.

Fig. 30.

In the Herbert Hospital the warming of the wards is effected by open fireplaces, whilst the subsidiary accommodation is warmed by hot-water pipes.

The wards are on the pavilion principle, with windows on opposite sides, the beds being placed between the windows. Each ward has its nurses' room and ward-scullery at one end, and its ablution- and bath-room and water-closets at the other. Every water-closet has a separate window, and the ventilation of these ward offices is most carefully cut off from that of the wards themselves. The walls are built hollow, and the windows are glazed with plate-glass, to save the heat, which the large extent of wall and window surface would otherwise allow to escape. The ward walls are of parian cement, the ward floors of oak, to be kept polished with beeswax. The ventilation of all the wards is as follows :—

First the windows open at top and bottom on both sides, then shafts, 14 by 14, pass up at each corner of the ward, to above the roof, to allow of the escape of foul air. Sherringham's ventilators are placed in each wall-space, between the windows, to admit fresh air when the windows are closed. For cold weather the principal engine of ventilation is the fireplace. There are two in each ward, placed in the centre line of the wards. The flue passes horizontally under the floor to a vertical flue in the wall. Fresh air is admitted by the side of the horizontal flue, through openings in the fire-clay sides of the fireplace, into the ward, by which means it becomes warmed ; so that each fireplace, when the fire is lighted. pours a continuous supply of fresh warmed air into the ward, and a great portion of the heat, which would otherwise pass into the chimney, is saved. The ablution-rooms and water-closets, as well as the lobbies which separate the ablution-rooms and water-closets from the wards, are each warmed by a separate service of coils heated by hot water from central boilers, and ventilated by fresh air admitted through the coils of these hot-water pipes. The vitiated air escapes by means of a separate shaft carried up through the roof from each ablution-room, water-closet, and lobby. The staircases, as well as the corridors separating the pavilions, are each warmed independently by hot-water coils, and ventilated by fresh air admitted through the coils. Each part of the hospital is thus independent of every other part for warming and ventilation.

In a new hospital at Glasgow and in one at Birmingham, Mr. Key propels the outer or town air through a specially arranged wet screen to purify it from fog and smoke, and then warms it. This system requires all apertures to be closed except those designed for the admission or removal of the air.

These instances sufficiently elucidate the general principles affecting combined warming and ventilation.

In all systems in which the walls derive their heat from the air of the room only they are necessarily somewhat colder than the air of the room. It has been mentioned that one of the main causes of the comfort of the open fireplace is that with it the air of the room derives its warmth from the walls which are warmed by the rays of the fire, and therefore that the walls are at least of the same temperature as the air; stoves and hot-water and steam pipes in a room also radiate a portion of their heat to the walls, those heated to a low temperature less effectively than those heated to a high temperature. Heating effected by warmed air can only be thoroughly comfortable when its use is combined with some plan of warming the floors, walls, and ceiling, so that their temperature may not be dependent on that of the air after it has entered the room.

In conclusion, no system of ventilation and warming in a large building or establishment can be satisfactorily conducted unless some person is charged with the duty of seeing that it is maintained at all times in effective action, on the principle of that adopted at the Opera in Vienna.

CHAPTER XIII.

CONDITIONS AFFECTING THE INTERNAL ARRANGEMENTS
OF BUILDINGS.

THE laws regulating the movement of air should govern the form of buildings.

Plans for houses, barracks, asylums, schools, and hospitals necessarily vary either with the wants of the individual or with public requirements, with the site, the aspect, and the cost. It is therefore impossible to do more than sum up the general principles which should be observed in a design. But in both private and public buildings architects should conform their architectual design to the internal requirements, and not, as is too often the case, make the internal arrangements conform to the design of the façade.

Fire-proof buildings are desirable ; a plaster covering to woodwork is a very good protection against fire. In the Communist riots in Paris, wood covered with plaster was charred, not consumed.

Fires frequently spread with rapidity in English houses because currents of air generated by a fire are enabled to pass up from one floor to another behind the boxing of windows, the battening on walls, the lath and plaster partitions, &c. The free passage of air-currents should be stopped between every room and the room under it and over it, by means of some substance difficult of combustion. There should also be a course of fire-clay slabs (Fig. 31), projecting at least

6 inches all round the walls of a room to receive the beams
of the floor above, and against which the plaster ceiling
should abut. Such arrangements
would greatly check the passage of
fire from storey to storey in a house.

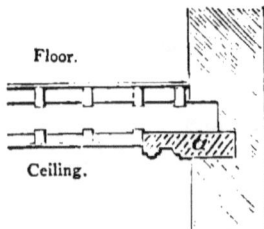

There should invariably be circulation of air round every building.

Back-to-back dwellings, as well as
houses and cottages built with in-
terlocking walls, are inadmissible, for
reasons already explained.

Floor.

Ceiling.

a. Solid slab of Fire-clay.

Fig. 31.

Dwellings should not be built over
stables, because it is impossible to keep the air of a stable
as pure as that of a living-room ; nor should dwellings be
placed over stores or shops, where matters liable to putrefy
are kept.

To guard against the liability of impurity in the ground on
which the house is built, when danger is apprehended from the
presence of decaying organic matter, the most healthy plan
would be to build the house over an arched floor raised above
the ground-level, the space under the floor being covered with
concrete and open to the air on all sides.

Basements are sometimes used in towns as dwellings : they
are undesirable. There should always be an area between
the basement and the street in houses situated in towns, to
prevent emanations passing into the houses through the earth
from defective gas-mains and sewers. Where cellars are provided underground, means should be taken to cut them off
from the ground-air.

Pure, dry air of the required temperature should pervade
every part of a dwelling. The form of the interior of a
dwelling should be such as to ensure this, and at the same
time prevent stagnation of air. In cold and temperate climates
when the weather is warm, and in hot climates always, change
of air largely depends on open windows and doors ; conse-

quently the relative position of windows and doors should be selected so as to enable the air of a room to be thoroughly changed when they are opened.

Lofty rooms are advisable, because the height facilitates the change of air without draught, provided the windows or other means of outlet for air be arranged so as to prevent stagnation near the ceiling, and so that impure and heated air which has risen to the upper part of the room may be removed rapidly. If time be afforded for its cooling, the impurities will fall and mix again with the air of the room. A lofty room, if warmed and provided with adequate means of change of air at the upper part, is more comfortable and healthy than a low room. In new houses or cottages, a height of 10 feet should be the minimum height for the best rooms, and no room in a cottage should be under an average of 8 feet high.

An abundance of light, and in this country direct sunshine, is always necessary for maintaining purity of air. In a hot climate means must exist for intercepting the direct rays of the sun.

A dark house is an unhealthy house, an ill-aired house, and a dirty house. Therefore light should penetrate to every part. There should be no dark staircases, corridors, corners, or closets. Direct light by means of windows easily opened to the outer air is required to ensure the frequent renewal of the air. Staircases, if lighted by skylights, should have the skylights made of the lantern form, with side-lights to admit fresh air without admitting rain.

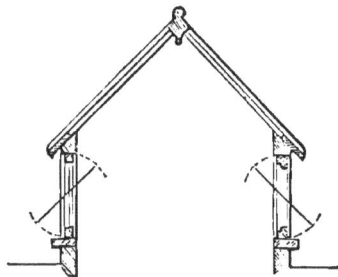

Fig. 32.

A hall or staircase in the centre of a house, carried up the whole height of the building, with light and ample ventilation by large windows at the top, forms a reservoir of air which may be kept fresh, and which materially assists in keeping pure the air of a house.

Every room in a building should have access to light and air, by means of a window in an outside wall. The rooms appropriated to removal of refuse, such as housemaids' closets and water-closets, require more light, and therefore proportionately larger windows, than other rooms. Store-closets should also have direct access to fresh air.

The larger the proportion which the area of surface occupied by a house bears to the number of occupants the better.

In towns where land is dear, and where a large number of persons are crowded on a given area, better ventilation and circulation of air may be obtained by placing dwellings in storeys one above the other, and leaving spaces between the buildings, instead of in one-storeyed buildings which would be too close together to allow of circulation of air round the building.

Under these circumstances, the height of a dwelling must be regulated with respect to its surroundings. That is to say, in the case of ordinary dwellings adjacent to each other, the distance apart of the dwellings should at least equal the height of the dwelling, so as to ensure adequate light and air in the lower floors. These considerations limit the number of storeys in houses of the better classes in towns, or in dwellings in which a large number of families are aggregated, in which case the arrangements should prevent a community of air throughout the building.

Basements should never be used for sleeping-rooms, nor indeed for human dwellings. They are always more or less liable to damp, stagnation of air, and deficiency of sunlight, and are well-known nurseries of disease.

The number of storeys in the case of barracks, workhouses, asylums, schools, and hospitals, where the conditions of the occupation entail a community of air throughout the building, must be more closely limited.

The following conditions afford a general idea of the considerations which affect these buildings.

In all such buildings, where a large number of human beings are to be lodged together, it is especially advisable, as a general principle, not to place anything which might injuriously affect the purity of the air in the same building with the inhabitants.

Kitchens, latrines, ablution-rooms, and baths, should therefore as far as possible be built away from them.

Buildings should be arranged in the simplest manner.

Squares with closed angles should be as far as possible avoided. The great object to be aimed at is to have free external ventilation all round the buildings; in temperate and cold climates to avoid a purely northern exposure; and in all climates, to ensure that sunlight may have access at some part of every day to all living-rooms. These conditions are essential to health. Free access of sunlight to a square is best obtained by placing two opposite angles of the square north and south.

If the administrative conditions allow of it, the simplest arrangement for such buildings is in a single line, lying north and south if possible, to allow the sun to shine on both sides of the range every day. The line may be divided into separate blocks for facility of passing across it at different points.

No part of any asylum, workhouse, or barrack, whether for sick or healthy men, should be placed close to the boundary walls. There should be always intervening space sufficient to ensure thorough ventilation round the buildings between them and the wall, and to prevent the ventilation from being injuriously affected by buildings belonging to the adjacent population placed close to the walls. Latrines, cook-houses, stores, and other similar buildings, can be placed between the building and the wall, but the arrangement should be such as not to interfere with the external ventilation of the building.

All populous buildings where there is an intercommunity of

air throughout the building, such as barracks, schools, work-houses, and asylums, are best constructed of only two storeys of living-rooms. Three storeys are not objectionable for healthy people, though very undesirable for sick. Four storeys should only be resorted to when, from restricted dimensions or from the form of the ground, it is absolutely necessary to adopt this number of floors.

Sleeping-rooms should invariably be raised above the ground level, with an air-space under the floors. Rooms for dry stores or for administration, or rooms occasionally occupied, such as libraries and reading-rooms, may be placed without detriment on the ground-floor, the sleeping-rooms being placed over them, when necessary. Barracks and hospitals have this in common—viz. that the sleeping-rooms are mainly also living-rooms ; they differ in this respect from asylums, schools, and workhouses. The following rule should apply uniformly to all buildings of the nature of an asylum, barrack, hospital, and workhouse—viz. such buildings should be subdivided into separate houses, without direct com-munication between the adjoining houses. To ensure this, the party walls between the houses should be carried above the roof.

Each house should further be divided up the middle by a wide roomy staircase, extending from the ground to the top floor, with a free ventilation through the roof. The staircase and passages should extend across the house from front to back, with windows on opposite sides for through light and ventilation. Besides affording means of access, the stairs and passages should be so constructed as to afford ventila-tion upwards between the two halves of the house, sufficient to prevent the atmosphere in rooms on opposite sides of the staircase and passages from intermingling.

There should be a unit of size for all asylums, hospitals, workhouses, and barrack-rooms, containing the principal rooms and their appurtenances, so that a building of any size, which has a definite object, may be constructed by simply increasing the number of such units.

The unit of construction for workhouses and asylums has not been laid down by any distinct authority; for barracks and hospitals the conditions laid down by the Barrack and Hospital Improvement Commission have been brought up to date in Hospital Construction[1]; and for prisons they have been stated by the late Sir J. Jebb.

The unit of construction for a school necessarily depends upon the nature of the school, and the number of classes into which it is divided.

These conditions vary in almost every case. It may however be assumed, that for school-rooms or lecture-rooms which are occupied for limited periods in the day-time, and thoroughly purified between times, a superficial area per pupil in the class-room of from 15 to 20 feet should be afforded, as a minimum, and a cubic space of not less than from 180 to 250 cubic feet. This, at 1,200 cubic feet per hour per occupant, would imply a renewal of the air of the room from five to seven times in the hour. But for a schoolroom occupied for limited periods, the windows being opened between times, 750 cubic feet per hour per occupant during school hours would probably suffice; reckoning after dark each candle as an occupant, and each gaslight as two candles. Windows in rooms where drawing or writing is carried on are best placed so as to be on the left hand of the student.

The conditions of school dormitories would follow those in barracks and workhouse wards which have been already alluded to in a previous chapter.

A short account of how the unit of accommodation for barracks was arrived at will explain the application of the general principle to a special case.

The volume of fresh air necessary for a barrack-room and the cubic space required for supplying this without draught

[1] Published by the University Press, Oxford.

was ascertained ; the number of occupants for the barrack-room was fixed at from twenty to thirty beds, the beds being arranged with their heads to the walls on opposite sides of the room.

There should be about half as many windows as there are beds in the room : they should be on opposite sides of the room ; they should be carried up to within a few inches of the ceiling, and be hung so that both upper and lower sashes can be opened or shut.

In no case should there be more than two rows of beds between the opposite windows. This rule holds good in all climates, but especially in hot climates.

Assuming a fixed floor-space, the width of the room must be such as to secure an adequate distance between the beds.

There is a minimum limit of width which depends on the question of convenience. Nineteen or twenty feet would be a good width for a barrack-room in this climate, but in hot climates a larger floor-space would be required, which entails additional width and length. The width above mentioned would allow space for tables and forms when the beds are down, and would allow about 11 or 12 feet between the opposite beds during the day, when the bedsteads are turned up.

Barrack-rooms should never be less than 12 feet high.

A room 20 feet wide and 12 feet high, with 5-feet bed-spaces along the walls, would give the regulation amount of 600 cubic feet and 50 feet of floor-space per bed. If the height of the room is less than 12 feet, it would be better to make up the unit of cubic space by increasing the bed-space along the walls, rather than by increasing the width of the room.

All sleeping-rooms should have ceilings. The space in the slope of the roof should not be taken into barrack-rooms any more than into the rooms of ordinary dwelling-houses. That space, if in the room, should always be ventilated ; otherwise it affords facilities for the impure air to stagnate, cool, and remix with the air of the room.

The fireplace should be placed in the side wall in the centre of the length of one side of the room, and should be constructed to warm part of the air admitted for ventilation. If the room were constructed for thirty beds, two fireplaces would probably be required; in which case both should admit warmed fresh air and be placed on the same side of the room, to diminish the probability of their smoking, and to assist ventilation.

Each barrack-room should have a room adjacent for a non-commissioned officer, opening from the landing or passage ; and either at the entrance, or at the further end opposite the entrance, there should be a well-lighted and well-ventilated room, with ablution-basins and a bath. There should be in addition a well-lighted water-closet with lifting seat to act as a urinal for night use, or else a place for a urine tub on the landing.

Hospitals require especial care, because the sick are more easily affected by insanitary conditions than persons in health.

The unit of hospital construction is the ward, with its ward offices.

The area of the ward depends on the floor-space allotted to the patients, to which allusion has already been made, and varies with the climate and with the object of the hospital. The necessary distance between the beds largely regulates the other dimensions. For general hospitals a bed-space of 7 or 8 feet is considered a minimum ; but in fever hospitals and surgical wards with bad cases as much as 12 to 15 feet for the bed-space has been adopted. It is essential that the windows should be opposite, and, in order that they may act efficiently for changing the air, that they shall not be too great a distance apart. The minimum breadth which combines convenience with this condition is 24 feet. The beds would stand between the windows.

The ward offices are of two kinds.

1. Those which are necessary for facilitating the nursing

and administration of the wards, as the nurse's room and room for medical man and ward scullery.

2. Those which are required for the direct use of the sick, so as to prevent any unnecessary processes of the patients taking place in the ward ; as, for instance, the ablution-room, the bath-room, the water-closets, urinals, and sinks for emptying foul slops. There should, in addition to the bath-room here mentioned, be a general bathing establishment attached to every hospital, with hot, cold, vapour, sulphur, medicated, shower, and douche baths. Separate water-closets are required for the nursing staff.

Hot and cold water should be laid on to all ward offices in which the use of either is constantly required, in order to economise labour in the current working of the hospital.

In large hospitals the nurses have sleeping, sitting, and dining rooms apart from their wards. In small hospitals the nurse's room should be light, airy, and well ventilated, close to the ward door, with a window looking into the ward. The ward scullery attached to each ward should be so placed as to be under her eye. There should be no dark corners in the scullery, and it should have ample window-space.

There should be provided, adjacent to the scullery or nurse's room, well-lighted rooms for linen, stores, or patients' clothes, and a hot closet for airing clean towels and sheets.

The ward offices of the second class ought to be as near as possible to the ward, but cut off from it by a lobby, with doors at each end, and windows on each side, and with separate ventilation and warming, so as to prevent the possibility of foul air passing from the ward offices into the wards. These offices are often conveniently placed at the end of the ward, farthest from the entrance and nurse's room, and distributed at each side, so as to enable the ward to have an end window.

The ablution-room should contain a small bath-room with one fixed bath, supplied with hot and cold water, and space for moveable bath with means of filling and emptying. A fixed terracotta bath when once warmed has the advantage of retaining the heat longer than a bath of almost any other material, and of being always cleanly, but it absorbs a great deal of heat at first. Hence, when the bath is consecutively used, it is the best material ; but if the bath is seldom used, then copper is better, or polished French metal.

The water-closets should never be placed against an inner wall, but always against an outer wall of the compartment in which they are situated, and the soil-pipe should be carried down outside, but protected from frost.

Walls of ablution-rooms and water-closets should be

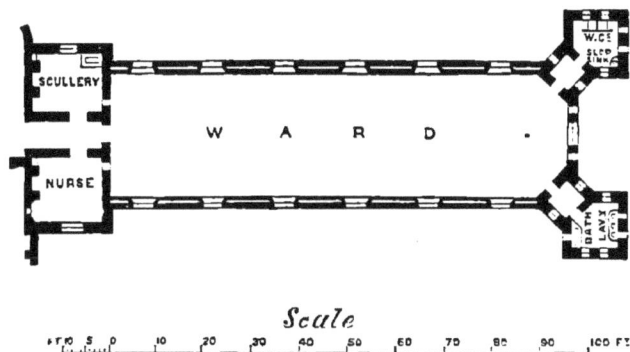

Scale

Fig. 33.

covered with white glazed tile, slate enamelled or plain, or trowelled cement ; plaster is not a good covering for them on account of their liability to be splashed, and of the necessity for the walls to be frequently washed down.

The ablution-room and water-closets should have plenty of windows opening to the outer air. They should have shafts carried up to above the roof, to carry off the foul air, and ventilated openings to admit fresh air independently of the windows, and warmed air should be supplied to them independently both of the wards and of the lobbies which cut them off from the wards, which latter should also be carefully ventilated and warmed separately.

These ward offices will vary but little with the size of the ward ; that is to say, a ward of twenty beds will require nearly as large ward offices as a ward of thirty-two beds. For instance, three water-closets per ward will suffice for a ward of thirty-two beds, but two at least will be required for even a twelve-bed ward. The superficial area to be added in the wards of thirty-two beds for these appliances would be about 30 square feet per bed, whereas in wards of twenty beds each it would come to nearly 50 square feet per bed.

To each large ward a subsidiary small ward for one or two beds, to receive bad cases, should be attached. These, with their ward offices, form a small hospital, which may be increased to any required size by the addition of similar units.

The principles upon which these units of ward construction, or, as they are generally termed, pavilions, should be added, are as follows :—

1. There should be free circulation of air between the pavilions.

2. The space between the pavilions should be exposed to sunshine, and the sunshine should fall on all the windows in the course of the day, for which purpose it is desirable that the pavilions should be placed on a north and south line.

3. The distance between adjacent pavilions should not be less than twice the height of the pavilion reckoned from the floors of the ground-floor ward. This is the smallest width between pavilions which will prevent the lower wards from being gloomy in this climate; and where from local conditions there is not a free movement of air round the buildings this distance should be increased.

4. The arrangement of the pavilions should be such as to allow of convenient covered communication between the wards, without interfering with the light and ventilation ; and therefore the connection between adjacent pavilions should be on the ground-floor only, or in the basement, and

the top of the covered corridor uniting the ends of pavilions should not be carried above the ceiling of the ground-floor wards. Indeed, whilst it is necessary to make the ground-floor wards from 12 to 14 or 15 feet high, it would be unnecessary for purposes of communication to give the corridor a greater height than 8 or 9 feet; there is however this consideration, that if the top of the corridor is made level with the ward floors of upstairs wards, it affords a convenient terrace on to which the beds of patients can be wheeled, so as to allow them to lie in the open air. Each block of wards—that is, each pavilion—should have its own staircase.

5. No ward should be so placed as to form a passage-room to other wards.

6. As a general rule, there should not be more than two floors of wards in a pavilion ; and the staircase leading to the upper floor should have an entrance separate from the entrance to the lower floor. Because, when two wards open into a common staircase, there is, with every care, to some extent a community of ventilation. When there are three or four wards one over the other, if they are all accommodated by one staircase, it becomes a powerful shaft for drawing up to its upper part the impure air of the lower wards, which is then liable to penetrate into the upper wards. Similarly, heated impure air from the windows of the lower wards has occasionally a tendency to pass into the windows of the wards above. A basement under the wards is of great advantage for keeping the ward dry and cutting off the ground air ; but they should not be used for any purpose, such as cooking, from which smells could penetrate into the wards, or for the storage of articles liable to decay, or for coal. When possible, it is best not to continue the staircase into the basement.

7. There is a limit to the numbers which should be congregated under one roof. This limit will depend very much on the nature of the cases. A careful consideration of the expe-

rience of military hospitals, into which many slight cases are received, led to the conclusion that 136 cases should be the largest number placed in one double pavilion, divided into two equal halves in such a way as to cut off by through ventilation the communication between the two. halves. In town hospitals, where the cases are of a more severe character, a similar double pavilion should probably not contain above eighty to a hundred beds.

The passage lobbies and staircases connecting the two halves of a double pavilion should be well lighted by large windows, and provided with ample ventilation direct from the open air ; in cold weather they should be supplied with warmth and fresh warm air independently of the wards and ward offices.

The size of any given hospital ought not to be determined by increasing the number of beds in any one building, but by increasing the number of units, each containing the numbers of beds mentioned ; and the extent to which these units should be multiplied might, if the units have been properly constructed and arranged, be determined not so much by the number of patients as by considerations of economy in administering the hospital.

To each of the larger wards it is necessary to attach a ward of one or two beds for special cases ; the number of these should be limited, to economise labour in nursing ; and their position must be adapted to suit the arrangements of the principal wards, and to afford easy supervision by the nurses.

It is moreover desirable that if convalescent patients remain in the hospital they should have rooms in which they can dine and spend the day apart from the other sick ; the situation of these rooms should be such as not to interfere with the light and air of the wards. This class of patients also requires a chapel. It is, however, a subject for consideration whether, as a rule, patients who are able to move about in this way should not be passed into convalescent institutions, instead of being retained in general hospitals.

All these arrangements must be made subservient to the broad general principle of giving air and light to the wards.

The corridors connecting the units may, in a warm climate, consist of an open arcade; in our climate in cold weather a closed corridor may be necessary: closed corridors should be lighted by windows on both sides, capable of opening wide, or of being removed altogether in warm weather; the corridors should be cut off from the adjacent pavilions by swing doors and be provided with separate means of ventilation, as well as with an independent supply of fresh warmed air in cold weather.

These arrangements prevent draughts, and cause the corridors, lobbies, and staircases to be the means of effectually cutting off the ventilation of one pavilion from that of another.

The staircases for patients should be broad and easy; the rise of each step should not exceed 4 inches in height, and the tread should be at least 1 foot in width; there should be a handrail on each side, and a landing after every six or eight steps.

The considerations which these data suggest are equally applicable, with modifications suited to the special case, to all buildings in which large numbers are congregated.

The sanitary condition of dwellings which are built for a special purpose, such as barracks, workhouses, hospitals, or asylums, can be easily controlled. There the occupation is continually of the same description. But in ordinary residences, the conditions are subject to change from the necessities or caprice of the occupants. In the houses of the rich, inconveniences arise from rooms built for everyday life being occasionally used for the reception of large numbers of people. Thus a dining-room which would be supplied with adequate fresh air for twelve people is sometimes used to contain thirty to dine, in addition to servants and numerous lights. Drawing-rooms adapted for a small number are often filled so full as barely to afford 3 square feet of surface per occupant, and are lighted by numerous gaslights or candles.

Rooms intended to be frequently applied to receptions or dinners, either in private houses, hotels, or municipal buildings, should be provided with permanent arrangements for the renewal of the air on a sufficient scale.

The heat which is generated by persons and lights will suffice in this country for affording a strong upward current in shafts to ensure the removal of air, provided the shafts are properly proportioned, and provided adequate inlets for fresh air be supplied ; and if the shaft terminates above the roof the movement of the external atmosphere will assist the removal. Dinners and receptions are limited in duration, hence it may be assumed that 750 cubic feet of air per hour supplied for each guest, servant, and candle, reckoning a gas-burner as two candles, would suffice. When arrangements for the removal of vitiated air and the inflow of fresh air have not been provided in the building, temporary arrangements may be resorted to, such as have been proposed by M. Joly of Paris and Mr. Verity in England ; viz. by means of a small fan worked by hand or by water power, which forces air into or extracts it from the rooms through a perforated india-rubber or other pipe laid temporarily along the cornice in a convenient situation. But such an arrangement is only a makeshift to remedy defects in the original construction of the building.

Artificial Lighting.

The impurity of the air caused by artificial light is very serious. Lamps, candles, and gaslights, each consume the oxygen and produce carbonic acid.

An oil-lamp with a moderately good wick burns about 154 grains of oil per hour, consumes the oxygen of about 3·2 cubic feet of air, and produces a little more than $\frac{1}{2}$ a cubic foot of carbonic acid ; 1 lb. of oil demands from 140 to 160 cubic feet of air for complete combustion.

A tallow candle of six to the lb. burns about 170 grains per hour, consuming the oxygen of about 4 cubic feet of air ; 1 lb. of tallow requires about 170 feet of air for combustion.

Coal-gas consists of olefiant gas and analogous hydro-carbons and hydrocarbon vapours, all of which contribute to its illuminating properties. It also contains hydrogen and marsh-gas, and in addition carbonic oxide, carbonic anhydride, sulphuretted hydrogen, and other sulphur compounds ; these latter are impurities.

The following table (from Dr. Tidy's Handbook of Chemistry) shows the composition of gas from cannel and from common coal :—

	Illuminating power compared to sperm candle burning 120 grains per hour, the gas burning 5 cubic feet.	Composition in 100 volumes.					
		Hydrogen, H.	Marsh Gas, CH_4.	Carbonic Oxide, CO.	Heavy Hydrocarbons, $(C_nH_n)_x$.	Equal to Olefiant Gas, C_2H_4.	Nitrogen, Oxygen, and Carbonic Acid.
Cannel gas .	34.4	25.82	31.20	7.85	13.06	(22.08)	2.07
Coal gas .	13.0	47.60	41.53	7.82	3.05	(6.97)	...

One cubic foot of coal-gas will (according to the quality of the gas) unite with from ·9 to 1·64 cubic feet of oxygen, and produces on an average 2 cubic feet of carbonic acid, and from ·2 to ·5 grains of sulphurous acid. In other words, 1 cubic foot of gas will destroy the entire oxygen of about 8 cubic feet of air.

The presence of 1 per cent. of carbonic acid in gas is said to decrease the light 6 per cent. To obtain a maximum light from any flame, the supply of air must not be excessive, otherwise the carbon particles are consumed before they are sufficiently heated to emit light, and the excess of atmospheric nitrogen serves to cool the flame and decrease its illuminating power. If, on the other hand, the supply of air is too limited, the carbon passes off unburnt, and the flame becomes smoky.

When gas is only partly burnt in a room, nitrogen, water, carbonic acid, carbonic oxide, sulphurous acid, and other impurities may escape into and vitiate the air.

It is most important to the purity of the air of a room that the products of the combustion of gas should not mingle with the air. Several forms of lights have been designed for this purpose, but as a rule they do not entirely fulfil their object, for in many cases they conflict with the other arrangements for ventilation. Thus, if a sun-light is placed in the ceiling, its proper action is to carry off a large volume of air from near the ceiling ; an open fireplace in the room draws a large volume of air towards itself. The currents conflict, and the fumes from the sun-light are liable to be drawn into the room.

In the case of a gas globe lamp suspended in the middle of the room, supplied with fresh air from the room, with a pipe to carry off the fumes leading from the light up to the ceiling, and along the ceiling into the chimney, the smallness of the pipe, the bend, and the horizontal length, all contribute a large amount of friction to diminish the draught in the tube ; in rooms where there is no arrangement for replacing the air removed by an open fire, cross currents will sometimes prevail in the chimney-flue, especially if it is large. Whenever the draught in the tube is sluggish, the fire in the chimney draws down some of the fumes directly towards the fire ; consequently it is very rare that, even with this class of burner, the fumes of the gas are removed.

The only safe plan is to place the gas-burners in a globe entirely cut off from the room, supplied with fresh air directly from the open air, the fumes being also carried directly into the open air. Such an arrangement is very simple in the case of an outside wall, and this is the only system which will keep the air of a room free from the fumes of gas. By the arrangement in Fig. 34 the fresh air is supplied through the grating to the globe from under the cap by means of the outer tube C, and the heated fumes pass up and away through the inner tube. When wind blows on or across the opening and the pressure is necessarily equal on both tubes, if the joints in the room are all air-tight the flame is not affected by wind. The incandescent gas-light, besides affording a brilliant light, consumes many of the deleterious products of an ordinary gas-burner; but it consumes the oxygen of the room, and it may emit other products special to the heating of the mantle.

The incandescent electric-light relieves us from the sanitary

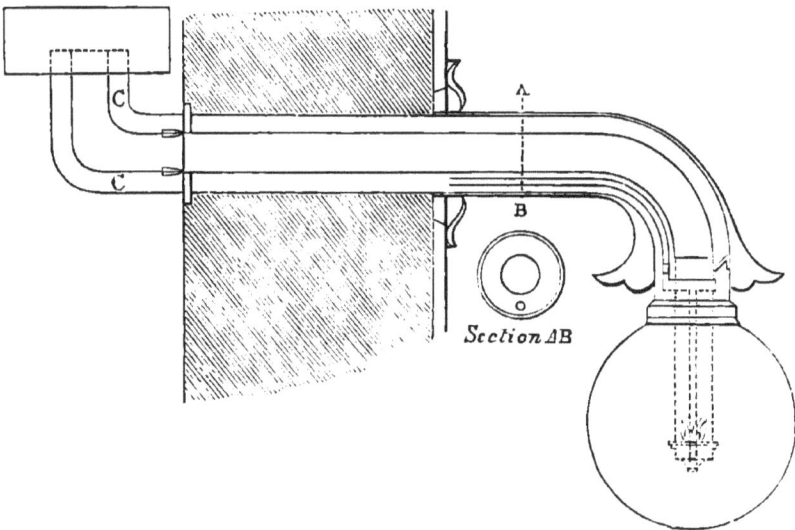

Fig. 34. Ventilated Gaslight.

defects experienced with other forms of lighting; but the electric arc-light consumes oxygen from the surrounding air, disengages nitric acid, and is not applicable to lighting an ordinary living-room.

Workshops.

In workshops the purity of air should be maintained by an adequate removal of vitiated air, and a supply of fresh air at the temperature which may be either necessary for comfort or

N

required in the processes carried on in the workshop. With
adequate renewal of air a temperature of from 75° to 78° or
higher can be easily borne, whereas without such renewal of
air lower temperatures soon become oppressive. In workshops where the processes carried on occasion fumes, steam,
or a large amount of dust, special extraction should be
arranged by means of fans creating rapid currents of air
to carry off at once the steam, fumes, or dust, and to prevent its mixing with the air of the room.

Stables.

It has not yet been ascertained how much fresh air
is required to keep a horse in health. Such an inquiry,
although of great value when warmth has to be combined
with ventilation, is of little comparative importance as applied
to stables, because the horse is not an exotic animal requiring
artificial warmth. He is taken from a perfectly open-air life,
with its vicissitudes of weather and temperature, to be confined,
more or less, in a stable for purposes quite apart from his
health. The question is, how to subject the horse to the
captivity he has to undergo in serving man, without injuring
him in his health and strength.

Animal life is most perfectly developed, and its functions
are most perfectly performed, under the conditions of free
diffusion of the atmosphere, including absence of stagnation,
abundance of light, good drainage, absence of nuisance, and
sufficient space to live in.

These are the conditions (besides of course food and drink)
which nature requires for the horse.

Good stable ventilation includes the other conditions, because
if the stable is filthy or ill-drained, or the ground saturated
with putrid urine, it must be obvious that no amount of fresh
air passing through the stable will keep it sweet and wholesome. Any amount of fresh air coming in will immediately be
tainted by filth which has already collected there.

Ventilator to each
box 3.0' long

Fan light
over door

Ventilating
course under
eaves

Continuous course
of air bricks

Air brick 9x6"

Section of a Loose Box 17ft. × 12ft.

Louvred Vent'r
the whole length

Continuous Skylight
on North side

Continuous
Course of airbrick
under eaves

Continuous
Course of airbricks
under eaves

Litter
shed

Continuous course
of air bricks

Air brick 9x6"

Section of Stable.

Elevation of Stable.

Windows over every Stall

Air inlet between every two stalls

6.6 × 6.6

Upper half sliding doors

Litter Shed

Litter Shed

paved

Shallow surface drain

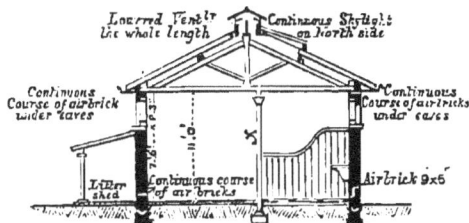

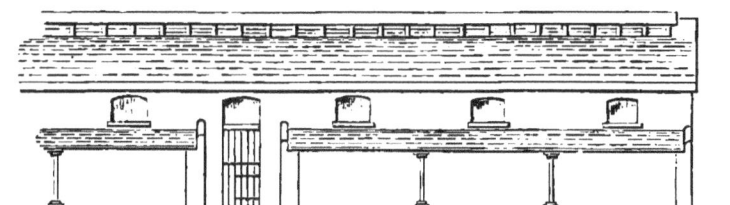

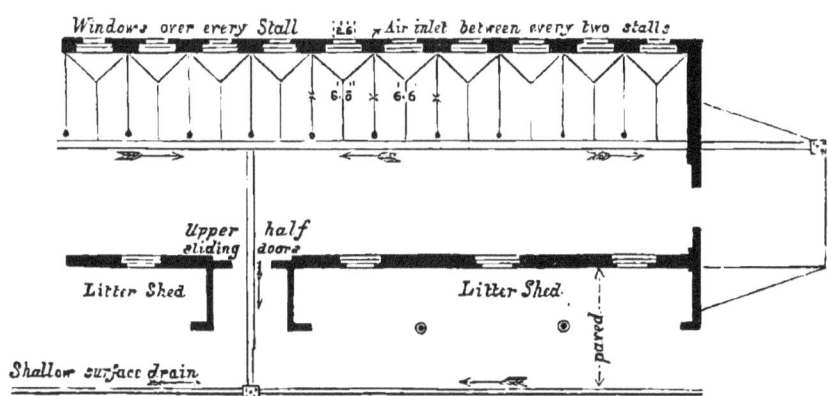

Fig. 35. Plan of Stable.

N 2

Again, if a stable be ever so clean or well drained, it will never be well ventilated without perfect freedom of movement of air through every part of it, together with free ingress and egress of air, so provided as to prevent hurtful blasts falling on the horses.

A fundamental requirement in all stables is paving of such a character as to wear well, not to become slippery, to be water-tight, and to be easily cleansed.

Another fundamental requirement is good stable drainage.

Surface drainage is the only kind of drainage applicable to the interior of stables.

The drains, like the stable floors, should be impervious to moisture. They should always be made of smooth material, with as few joints as possible, be carefully laid, having a shallow saucer-shaped section, and with as rapid an incline as it is possible to obtain. They should pass behind the line of stalls, and be conducted in as straight a line and by as short a course as possible to the outside of the stable, where they should be discharged into an underground drain, over a trapped gulley-grating, placed at a distance of some feet from the stable wall, so as to prevent effluvia returning, and to prevent dung and straw from entering the drain.

The most scrupulous cleanliness of the surface of the stable should be enforced.

The great principle which ought to be kept in view in stables is to have the air moving freely through every part of them, above and around the horses when they are standing, and in all the angles between the floor and walls when the horses are lying down, and every horse should have sufficient ventilation for himself without being obliged to breathe the foul air of his neighbours. These conditions would be most completely obtained in an open shed, such as is used for stabling horses in warm climates, and the nearer we can approach to this construction, keeping in view the necessity for protecting horses in this climate, while at rest, from extreme cold and cold blasts of wind, the healthier will be the stable.

That form of construction which affords the maximum of facility for obtaining a free moving atmosphere throughout the body of the stable is the open roof with ridge ventilation carried all the way along.

Where the roof of the stable is not open, but flat and impervious, the distance between the effective ventilating openings, whether windows or other apertures, corresponds of course to the breadth of the stable. But with an open roof and ridge ventilation the distance is reduced to one-half, while the difference of height above the ground between the ridge opening and the side windows ensures a far more certain and continuous movement of the air than could by possibility take place with side windows, unless a high wind were blowing. Therefore a stable with ridge ventilation is the most healthy stable.

A flat impervious roof, a hay-loft, or a barrack-room over a stable increases the difficulties of ventilation.

In so far then as concerns the general movement and renewal of the mass of air in a stable, the form of construction which effects this most easily and efficiently is an open-roofed stable, with ventilation along the ridge, swing windows along the sides, and a continuous inlet for fresh air under the eaves made of perforated brick, so arranged as to throw the entering currents up towards the roof.

A great incidental advantage of the open roof should not be overlooked, and that is the facility with which it enables the stable to be thoroughly well lighted. Light is in its place as essential to health as air, and moreover, when introduced vertically from the roof, it enables the state of cleanliness of the stable to be seen at once.

Besides providing for free movement of the mass of air within the stable, it is necessary in all stables, but in some much more than in others, to supply fresh air near the ground-level at the head of each stall, so that the horse may have fresh air to breathe when he is lying down.

The reason of this necessity is that in all stables the stratum of air next the floor-level is the most impure, and will always be the most impure under any improved conditions of drainage and paving. Besides this, the horse, in lying down, places his head close to the angle between the floor and the wall where the air is stagnant.

It follows from what has been said that the easiest and most efficiently ventilated stable is the open roof partially glazed, with ridge ventilation all along, ventilation at the eaves, a swing window for every stall, and the horses' heads turned outwards, with a proper air-brick in the outer wall, introduced 6 inches from the ground between every two stalls.

With these conditions of ventilation the Barrack and Hospital Improvement Commission stated that each cavalry horse should have 1,600 cubic feet and 100 superficial feet of space.

General Morin states that the cubic space allowed in French cavalry stables is 1,750 cubic feet per horse, and he lays it down that 7,000 cubic feet is the volume of fresh air which should be supplied, or of vitiated air which should be removed, per horse per hour.

CHAPTER XIV.

CONDITIONS AFFECTING MATERIALS AND DETAILS
OF CONSTRUCTION.

IN every dwelling dryness is an essential of health. In this climate it is necessary to provide for warmth. In a hot climate coolness is sought. In a hot climate, or for hot weather, the roofs and walls should be of such construction as to prevent the temperature inside a house being raised by the heat of the sun. For cold weather, on the other hand, the roof and walls must keep in the heat, so as to maintain the air of the house at a higher temperature than the outside air.

The materials and constructional arrangements best adapted to keep warmth in a house will to a considerable extent be effective for keeping out the heat.

The temperature which can be maintained in a house will greatly depend on the construction of the walls, and on the materials of which they are composed. Materials differ greatly in their power of allowing heat to pass through them.

The following examples, showing for different materials the units of heat transmitted per square foot per hour by a plate 1 inch thick, the two surfaces differing in temperature 1° Fahrenheit, illustrate their relative advantages in this respect.

Marble, grey fine-grained 28
Do. white, coarse-grained . . 22
Stone—ordinary freestone . . . 13.68
Glass 6.6
Brickwork 4.83
Plaster 3.86
Brickdust 1.33
Chalk powdered 87
Fir planks 1.37

Increased conductivity of heat in a material must be counteracted by increased thickness. All ordinary wall materials admit of a greater or less change of air through the material itself, depending upon the extent to which the material is porous.

The porosity of a material is shown by its power to absorb water. The following [1] is the percentage of its own weight of water which each of the materials mentioned below has been found to absorb—

Bricks.	per cent.	*Stones.*	per cent.
Malm Cutters	22	Good Granite	½
Malm Bright Stock	22	Indifferent Do.	1
Malm Seconds	20	Bad specimen Do.	3
Brown Paviors	17	Trap and Basalt	a trace
Hard Paviors	9½	Sandstone—	
Common Grey Stock	10½	Craigleith	8
Hard Do.	7½	Parkspring	8
Washed Hard Stocks. . . .	4½	Mansfield	10.4
Staffordshire—		Hassock (very bad quality).	20
Common Blue	6.5	Limestone—	
Dressed Do.	2.3	Marble	a trace
Brown glazed brick	8.6	Portland	13.5
		Ancaster	16.6
		Bath	17
		Chilmark	8.6
		Kent Rag	1⅛
		Ransome Artificial stone . . .	12

From this it appears that brick and stone walls, being always more or less porous, must admit, as already mentioned, of a considerable spontaneous change of air when dry.

[1] See Notes on Building Construction, Science and Art Department, 1879.

Porous walls tend to absorb the moisture given out in breathing, or in the combustion of lights. In this process walls absorb organic and other impurities, which after a time decay, and the wall may become a source of impurity for the air. When a wall is damp, change of air can no longer go on through it. The evaporation from the wall cools down the temperature. Damp in walls is considered to be a cause of fever, especially in warm climates. The damp wall, whilst it checks· the passage of the air, is cold, and consequently occasions a rapid radiation of heat from persons sitting within its influence. On the other hand, in hot and dry climates, wet mats are often hung over all openings, as a means of cooling the air without injury to health. These represent absolutely pervious walls, admitting of a rapid change of air.

In this climate, damp walls, besides being unhealthy, are uneconomical. They cause a great absorption of heat by the evaporation of the moisture from the surface. New walls are always damp. The quantity of water which will be contained in a new wall is very remarkable. Suppose that 100,000 bricks are used for a building, each weighing seven pounds; a good brick can suck up from 10 to 20 per cent. of its weight in water, but assume 7 per cent. as what gets into it by the manipulations of the bricklayer. Also assume that the same amount of water is contained in the mortar, a quantity certainly much understated; the mortar forms about one-fifth of the walls; thus nearly 100,000 pounds of water, equal to 10,000 gallons, may be assumed to be put in the walls in the process of building, and which must be removed from the walls of the house before it becomes habitable. This water must be removed by evaporation into and by the air. The capacity of the air for receiving water depends on the different tension of the vapour at different temperatures, on the quantity of water already contained in the air as it flows over this moist surface, and finally on the velocity of that air. Assume the average temperature of the year to be about 50° Fahrenheit, and the average hygrometric condition of the air to be 75 per cent. of its full saturation. At the temperature named one cubic foot of air can take up four grains of water in the shape of vapour, but as it contains already 75 per cent. of these four grains, which amounts to three grains, it can only take up one additional grain. As often then as one grain is contained in the 10,000 gallons of water mentioned above, as many cubic feet of air must come in contact with the new walls, and become saturated with the water contained in them ; or, about 700,000,000 cubic feet of air are required to dry the building in question. Therefore the drying of a building will be best effected by passing a large volume of air through it, and air at a higher temperature, and therefore of a greater hygrometric capacity than the outer air, will effect this object most rapidly.

Until this damp has been expelled by fires or by time, the building should not be occupied ; when the walls have been dried inside, it is the proper function of the walls to prevent damp from entering from the outside. Damp should be precluded from rising into the wall from the ground by means of a damp course carried round in all the walls below the level of the lowest floor. The damp course may be of slate, asphalte, or glazed perforated bricks ; this latter form of damp course has the advantage of allowing air to penetrate on all sides uniformly under the basement floor.

Damp should be prevented from descending into the wall from above by an impervious coping or by eaves. Improperly laid copings will act as conductors of wet into the wall, rather than as protectors against it. Projecting eaves are advantageous ; the more they project, the greater protection do they afford to the surface of the walls. The rain which beats against a wall will be partly evaporated out again by the action of the air and sun, and partly drawn through it by capillary attraction ; and if the material is very porous, or the wall very thin, it may saturate the wall. The capillary action will be checked by joints in brick or stone work, and arrested by air-spaces in the wall or by the use of a hollow wall.

The heat generated in rooms passes away in cold weather through the floor, ceiling, and walls and windows into the open air or adjacent colder parts of the house, and the heat of the house is similarly constantly passing away into the open air through the walls and the windows and the roof.

The temperature of the basement floor when below the level of the adjacent surface will be practically that of the mean annual temperature, and will therefore occasion little loss of heat.

The loss of heat by walls varies in a direct ratio with the conducting power of the material of the walls, and with the difference of temperature between the inner and the outer surface of the wall, and it varies inversely with the thickness of the wall. The actual temperature of the surface of the wall is troublesome to ascertain: if instead of the temperature of the surface of the wall, the temperature of the air inside the building and outside be taken into account, the formula becomes somewhat more complicated, and varies in a direct ratio with the conducting and radiant power of the material of the wall, the loss from contact with air, the difference of temperature between the air inside and that outside, and in an inverse ratio with the thickness of the wall[1].

[1] If T = temperature of internal air,
T' = temperature of external air,
R = radiant power of the material,
A = loss by contact of air,
C = conducting power of the material,
$Q = R + A$.
U = units of heat per hour,
E = thickness of the wall in inches,

the rule is—

$$U = \frac{(A \times C \times Q) \times (T - T')}{\{C \times [2 \times A + R]\} + \{E \times A \times Q\}}.$$

From Box on Heat.

The loss of heat by a vertical wall from contact of cold air per square foot of area will be greater with a low wall than with a high wall. The reason is obvious; the cold air in immediate contact with the lower part of the warmer wall receives heat from it, and ascends, and at each successive gradation in height the difference of temperature between the air and the wall is decreased by these successive increments of heat, and the amount of heat given out by the wall to the air is thus progressively diminished as the air reaches the upper part of the wall, in a proportion dependent on the square root of the height.

Thus whilst a vertical plane 1 foot high would lose ·594.5 units of heat per square foot per hour for each 1° Fahrenheit difference in the temperature between the surface of the plane and the adjacent air, a vertical plane 10 feet high would lose ·4350 units of heat per square foot, one 40 feet high would lose ·3980, and one 100 feet high would lose ·3843 units per square foot.

The problem connected with the loss of heat by walls requires more space for its full discussion than a treatise of the nature of this one, limited to the general enunciation of principles, will admit of. For the special study of the question, Péclet, Balfour Stewart, Box on Heat, and other writers may be advantageously consulted. From the latter work, which is eminently practical in its character, the following table is extracted. The units of heat transmitted per square foot per hour by a plate 1 inch thick, the two surfaces differing in temperature 1° Fahrenheit, being as shown by Péclet's experiments previously alluded to—

for ordinary stone = 13·68,
for brickwork   = 4·83.

The table shows the loss of heat per square foot per hour by brick and stone walls, 40 feet high, in buildings where only one face is exposed, and for 1° difference of internal and external temperature.

Brickwork.		Stone.	
Thickness.	Units of Heat	Thickness.	Units of Heat.
brick. inches.		inches.	
$\frac{1}{2}$ = 4$\frac{1}{2}$	·371	6	·453
1 = 9	·275	12	·379
1$\frac{1}{2}$ = 14	·213	18	·324
2 = 18	·182	24	·284
3 = 27	·136	30	·257
4 = 36	·108	36	·228

The number of units of heat lost through a hollow wall, or wall with an air-space in the centre, is less than that through a solid wall. The units of heat, dissipated by the outer air, will be in a direct ratio to the difference between the temperature of the air-space in the wall and that of the outer air; whilst the effect of diminished thickness in the wall follows an inverse ratio somewhat less than that of the actual diminution of thickness of the wall. For instance, if the difference of temperature between the room and the outer air $= t$, and the temperature of the air-space be a mean between that of the room and the outer air, then the difference of temperature between the air-space and the outer air will equal $\frac{1}{2}$ t, and if further the thickness of the wall be E, and the air-space be in the middle of the wall, so that the thickness on each side of the air-space $= \frac{1}{2}$ E, the formula in the preceding note for ascertaining the value of the units of heat lost in each case, expressed briefly would assume the general form, in the case of the wall without an air-space, of $\dfrac{L\,t}{M+EN}$ and in the case of the wall with an air-space, of $\dfrac{L\,t}{2M+EN}$, or, if we assume from the previous table a wall of a room 9 inches thick, with a temperature in the room 2° above that outside, the loss of heat would be ·550 units per square foot per hour. If, on the other hand, two walls half a brick (or 4$\frac{1}{2}$ inches) thick each, were used on each side of an air-space instead, and the temperature of the intermediate air-space was 1° above the outer air, and 1° below the temperature of the air of the rooms, the loss of heat would be ·371 per square foot per hour, or about two-thirds of the heat lost by the solid wall.

The loss of heat, as well as the porosity of a wall, is influenced by wall coverings, such as stucco outside ; or plaster, cement, or papering inside.

Independently of the advantages which may be derived from the smaller conductivity, if any, of the covering material, each layer forms an additional break in the continuity of any one material, and lessens both the porosity and the loss of heat.

The best internal wall-surface for a dwelling would be an impervious polished surface, which would not absorb moisture from breathing, and which, on being washed with soap and water, and dried, would be made quite clean. This wall would not absorb organic matter, but when colder than the air of the room, the moisture from the breath, &c., would be condensed upon the wall.   To prevent this, either the temperature of the wall should exceed that of the air, or a larger volume of air would require to be passed through a room with impervious walls than a room with pervious walls.   In conveying warmth to rooms with impervious walls by an agency different from an open fire, it would be preferable to convey it in such a manner as to warm the walls, and warm the air of the room through their means.   An enamelled metallic wall surface, with a space between the surface and the brick or stone wall for the passage of warmed air, would effect this object.

Plaster, wood, paint, and varnish, all absorb the organic impurities given off by the body, and any plastered or papered room, after long occupation, acquires a peculiar smell.

In a discussion, in 1862, in the French Academy of Medicine, a case was mentioned in which an analysis had been made of the plaster of a hospital wall, and 46 per cent. of organic matter was found in the plaster. No doubt the expensive process which is sometimes termed enamelling the walls, which consists of painting and varnishing with repeated coats, somewhat in the manner adopted for painting the panels of carriages, would probably prove impervious for some time, but it would be expensive, and very liable to be scratched and damaged.

Parian cement polished is practically an impervious material, but it is costly; unless carefully applied, its appearance is unsatisfactory, and it can only be applied on brick or stone walls, and not on wood-work or partitions, because, being inelastic, it is liable to crack. The want of elasticity in Parian cement is unfavourable to its use in ceilings.

The numerous joints required for glazed bricks, or tiles, render the use of these questionable as wall linings. The cement of the joint being more or less porous is sooner or later discoloured. Moreover, cracks and joints are objectionable, as they get filled with impurities, and may even harbour insects.

In default of any impervious covering, the safest arrangement in hospital wards is plaster lime-whited or painted, which should be periodically scraped so as to remove the tainted surface, and be then again lime-whited or painted. In hospitals, of course, these arrangements require the wards to be periodically vacated; but this is of itself an advantage, because every ward should be left empty annually for a period.

When plaster is used, it is essential, for the reasons before mentioned, that at the expiration of a number of years, dependent upon the degree to which the room has been occupied, the whole outer coat of plaster should be removed from the walls and ceilings, and new plaster substituted. When walls are re-papered, the old paper should be invariably removed, as it is saturated with organic matter. In all places occupied by many persons, such as hospital wards, barracks, or asylums, the walls and ceilings should be quite plain, and free from all projections, angles, or ornaments which could catch or accumulate dust.

In connection with wall coverings, it is necessary to allude to the danger of poison in wall-paper or paint. Arsenic is the substance from which this danger is generally found to arise.

Green paper, as a rule, contains more arsenic than others; but colour in paper is no guarantee of freedom from arsenic. If not in the colour itself, it may still be in the mordant dyes or other material used in the manufacture of the paper. Arsenic in various combinations more or less dangerous is used in a great variety of colours. Many French greys and neutral tints, and some white papers, are as heavily loaded as green. Nothing but an examination of the individual sample of paper or colour will afford security against its presence.

The danger is in proportion to the quantity of arsenic or mineral poison used in the colouring matter of the paper; and in proportion to the facility with which it may be removed from the fabric, either as dust or gas.

Danger from arsenic as a colouring matter seems to depend in part upon the presence of size. Dr. Fleck showed by experiment that a mixture of arsenious acid and starch paste or other organic substance gives rise to the formation of arseniuretted hydrogen, but no arsenic could be detected in air which had been in contact with a mixture of arsenious acid and water without the presence of organic matter.

Arsenic is frequently present in distemper, which being mixed with size forms a direct combination of arsenic and organic matter, liable to give off arsenic under many circumstances; and in the case of damp walls it is there ready for the development of arseniuretted hydrogen.

Arsenic is a powerful antiseptic, and hence the danger of its being introduced into glue and size, because it is so effectual in preventing decomposition, and is free from smell; moreover, size is largely used for fixing colours. In purchasing wall-papers a guarantee should be required from the seller; but it is also desirable to have the specimen analysed.

Floors.

One of the most important conditions to be observed in the materials for floors in this climate is that they should not be cold to the feet, consequently wood floors are desirable, unless the tiled floor is arranged so as to be warmed on the old Roman plan—viz. by means of tiles laid with a hollow space or flues underneath, warmed by the smoke or heated gases from a furnace.

Concrete, cement, and stone and brick more or less permit the passage of damp ; therefore the floor, of whatever material it be, should always have an air-space under it so as to ensure dryness.

The basement floor of a house may have an important influence on the purity of air. The heat of the basement floor will be substantially that of the ground under it, and that temperature differs but little from the mean annual temperature of the ground. The basement will thus be cooler in summer and warmer in winter than the outside air. This warmth tends to cause the air to rise through the house, and hence in cold weather, if warmed fresh air is not otherwise provided in a house, and if there is a basement, the air will be liable to pass up from the basement and pervade the house. Consequently, in such a case, the purity of the air in the house will depend upon the purity of air in the basement.

The purity of air in the basement will depend upon the arrangements for keeping all refuse and objectionable things out of the basement itself, but it depends especially upon the ground-air in the subsoil under and around the house being entirely cut off from the basement. To secure this, a continuous bed of concrete, or a layer of asphalte, should be laid over the whole surface covered by the house, and sufficient areas should be carried below the level of the basement so as to cut the ground-air off by an open air-space from the adjacent soil.

In order further to diminish the liability of ground-air to penetrate into the basement, there should be an air-space between the ground and the floor of the basement.

The only case in which it is advisable to dispense with an air-space under the floor would be when the floor is of tiles or of wooden blocks laid on concrete and embedded in asphalte, to ensure dryness. A floor of wooden blocks laid on and bedded in asphalte combines dryness with warmth for the feet.

The floor of the basement, whether it be of wood, of stone, or of tile, should be from one foot to eighteen inches above the level of the surface of the ground around it : where the floor is boarded the joists supporting the floor should not be laid on the ground, but should have a space underneath, to which a free circulation of air should be admitted, by means of gratings communicating with the outer air.

Drains should not be carried under the basement-floor ; but where unavoidable special precautions should be adopted.

With wooden huts care should be taken to level the ground under the floor, and to allow of free circulation of air, either between each pair of joists, or, what is preferable, to raise the joists so as to allow free circulation under the whole floor.

Floors should be laid with close joints, and wooden floors should be tongued, so as to prevent dirt from falling through and accumulating under the floor, as such dirt is liable to putrefy.

The frequent saturation of wooden floors with water to keep them clean diffuses damp ; consequently a closely laid polished floor of hard wood possesses great sanitary advantages.

For hospitals the floor requires special considerations. Ward-floors should be as non-absorbent as possible, and for the sake of warmth to the feet they must in this country be of wood ; oak, or other close hard wood, with close joints, oiled and beeswaxed, and rubbed to a polish, makes a very good floor, and absorbs very little moisture. It is impossible to pay too much attention to the joints; they should be like those of the best parqueterie, affording no inlet for dirt to lodge; because the impurities which become lodged in the cracks of a hospital floor are eminently objectionable. There should be no saw-dust, or other organic matter subject to decay, under the floor. When one ward is placed above another, it is essential that the floor should be non-conducting of sound, and that it should be so formed as to prevent emanations from patients in the lower ward from passing into the upper wards. The floors of the Herbert Hospital are formed of concrete, sup-ported by iron joists, over which the oak boards are laid.

An economical and non-absorbent surface for the floor can be obtained by first laying rough deal boards and covering them with thin, closely-laid oak boards oiled and beeswaxed. These oak floors must be treated like the French parquet, by occasional frottage. A very good hospital floor is one which is oiled, lacquered, and polished, so as to resemble French polish. It is damp-rubbed and dry-rubbed every morning, which removes the dust. This wet and dry rubbing process of cleaning is far less laborious than either frottage or scrubbing, and completely removes the dust and freshens the ward in the morning. The only objection to this surface is its want of durability, and consequent necessity for periodical renewal. Both of the processes above mentioned render the floor non-absorbent, and both processes do away with the necessity of frequent scouring, which is objectionable from the quantity of damp it introduces into the ward. The French floor stands the most wear and tear, but must be rubbed periodically by a frotteur, which cleaning is more laborious than scrubbing. The daily cleaning of a beeswaxed floor and the removal of dust may be effected by wiping with a damp cloth wetted with warm water, and carefully rubbing with a dry cloth. Practically, with care, a well-laid oak floor, with a good beeswaxed surface, can always be kept clean and polished in this manner assisted by periodical frottage.

Roofs.

The outer covering of a house should be impervious to moisture from without ; but experience shows that it is unhealthy to live under a ceiling impervious to air. Air heated by contact with the human body carries up emanations which, when they rest upon a pervious ceiling, are retained there, whilst the moisture passes off through the ceiling ; on the other hand, if these emanations come in contact with an impervious ceiling, they are not absorbed, and may be again brought into circulation in the air of the room. Consequently, if circumstances render it necessary to have a ceiling impervious to air and moisture, this must be discounted by providing under such impervious covering additional facilities for change of air in the upper part of the room.

The ordinary lath and plaster ceiling, although a good non-conductor of heat, allows of a considerable passage of air and moisture. With a pressure obtained by a difference of temperature of $72°$ inside and $40°$ outside, the quantity of air which was found to pass through ordinary plaster was about 1·5 cubic feet per square foot of area per hour.

The nature of the roof must depend on the materials available. Thus cement, clay, tiles, wooden shingles, slate, iron, copper, lead, are used according to circumstances. If good conductors of heat are used to keep out wet, such as copper, iron, or slates, some non-conducting material is required underneath ; for instance, in the case of a slated roof, the slates should be laid on close boards covered with felt, to ensure the best protection against heat or cold.

The loss of heat through the roof will depend upon whether the rooms are ceiled, and upon the form and nature of the roof-covering. If there is a lath and plaster ceiling to the upper rooms, and an air-space between the ceiling and the roof, closed from the outer air, so as to prevent any rapid circulation of air, and if the roof be formed in the most approved manner in this country, viz. with close boards covered with felt under the slates or tiles, the loss of heat in winter, and the effect of heat in summer in raising the inner temperature, will be comparatively small.

If there is no ceiling, and if the roof be not carefully constructed, as above mentioned, the loss of heat will be very considerable. The loss of heat from glazed roofs, ceilings, and skylights, and from metal roofs, such as are used in railway stations and markets, is very considerable.

The air in contact with metal or thin glass exposed to cooling influences is under the most favourable condition for being cooled. Each layer as it is cooled falls down and is replaced by warm air, which undergoes the same process. This renders a space covered with a metal or glass roof without intermediate ceiling very difficult to warm. Therefore in halls or rooms lighted by a glass roof, or staircases lighted by skylights, it is essential for preserving the heat that there should be a second glass ceiling below the one exposed to the outer air ; and in cold weather it may be advisable to adopt special means to warm the intermediate space, if an equable temperature is sought to be maintained in the room at all times. This can be effected by hot water or steam-pipes, or where glazed ceilings are lighted by gas-lights above the lower ceiling of glass, the heat from the gas when lighted is sufficient to keep up the temperature.

In very hot weather, when it is desired to cool down the temperature of an iron or glass roof, it may be watered by jets from 8 or 9 o'clock in the morning till about 5 o'clock in the evening. The quantity of water would however be considerable, probably not less than about 25 gallons per hour per square of 100 feet. The practice of lime-whiting the roof, which is largely resorted to in some places, is a great protection against heat.

Windows should be carried up as near the ceiling as possible. In a low room this is essential, in order to freshen the

air. In a lofty room it is equally necessary, in order to prevent the warm impure air which has risen to the top, from
cooling, falling, and remixing with the air in the lower part of
the room.

Therefore if the tops of the windows are much below the
ceiling there should be openings above the windows for change
of air. Moreover, whilst it is an element of cheerfulness in
a room for the upper part of the windows to be near the ceiling, it is equally important that the sills of the windows should
be brought down near to the floor.

Similarly, it is an element of cheerfulness to splay the sides
of window openings.

The proportion of window surface to the cubic contents of
the room must to some extent depend on the climate and aspect.

In England, adequate light will generally not be obtained
with less than one square foot of window-surface to about 100
cubic feet of the contents of the room.   In hospital wards. one
square foot to 60 cubic feet of content is desirable.   This
should be a minimum allowance, and assumes that the windows in all cases are of clear glass ; but greater cheerfulness
will be secured by more light.   In climates with bright sunshine less light may be found sufficient.   Where light is abundant it can always be modified, when desirable, by sunshades
or blinds ; but if window openings are small, light cannot be
increased at will.

The amount of light afforded by a window is considerably
modified by the quality of glass.

In some recent experiments :—

Polished British plate glass, ¼ inch thick, intercepted 13 per cent. of the light.
36 oz. sheet glass . . . .   „   22   „   „
Cast plate glass, ¼ inch thick . . . .   „   30   „   „
Rolled plate glass, 4 corrugations in an inch „   53   „   „

Clear glass is thus of great importance, and the thicker it is,
consistent with clearness, the better, because thin glass allows
of a more rapid loss of heat.

Good glass is desirable, because dust adheres easily to bad glass, but not to good glass ; and the surface of bad glass is more or less rapidly rendered uneven by exposure to the atmosphere, whilst good glass is unaltered by long exposure. The quality of glass depends upon the admixture of the ingredients. The percentage composition of window glass is 66·37 of silica, 14·23 of soda, 11·86 of lime, 8·16 of alumina ; that of plate glass 73·5 of silica, 5·5 of potash, 12 of soda, 5·5 of lime, 3·5 of alumina. A soda glass is more fusible and more brilliant than a potash glass. But soda imparts a slightly green tinge to glass, which does not occur with potash. Lime diminishes the fusibility of the glass, imparts no colour, and increases its hardness and lustre.

The most convenient form of window for ventilation in ordinary dwelling-rooms in this country is the sash-window, opening top and bottom. This mode of construction assists ventilation in the manner already described by enabling a current to be maintained ; when both sashes are open fresh air enters below whilst the vitiated air of the room passes out at the top. The lower sash should have a broad bottom bar in order that the sash may be raised from one to two inches without admitting air at the bottom, but so as to allow of a narrow opening between the two middle bars through which air could flow in vertically upwards. Ventilation should not however be dependent only on windows. In hospitals, where the wards, and consequently the windows, are lofty, the lower part of the windows may be advantageously constructed of the sash form, whilst the upper part is hinged on a transom so as to open inwards, and thus facilitate the inflow and outflow of air, on the plan of a hopper ventilator.

In hot climates the French window is advantageous, as it enables the whole window opening to be utilised for supplying fresh air.

The loss of heat through windows amounts to that lost by radiation added to the loss of heat by contact with air.

It may be assumed with thin glass that the temperature of the outer surface of the glass is a mean between the temperature inside the room and that of the outer air.

For thin glass, adopting the previous notation in the note, page 187,

$$U = (R + A)(T - T^1).$$

With thick glass the conducting power of the material must also be taken into account, as in the case of a wall.

The loss of heat with double windows is much less than that with single windows, and they have the advantage not only of transmitting less heat, but, from the temperature of the inside glass being greater, less radiant heat is absorbed from the occupants of the room. Péclet found that the loss of heat in double windows increased somewhat with the distance apart of the inner and outer glass, owing probably to the greater facility for currents of air in the wider space between the glass.

Thus with an intermediate space between the windows of ·8 of an inch, the loss of heat of the single window to that of the double window was in the proportion of 1 : ·47; with a distance apart of 2 inches the proportion was as 1 : ·55; with a distance apart 2·8 inches, which is nearly what exists in practice, the proportion would probably be as 1 : ·6.

The proportionate loss of heat by walls, as compared with the loss of heat by windows, varies to some extent with the conditions of the room, i.e. the proportionate extent of wall exposed to the outer air; but with 14-inch brick walls, and an assumed internal temperature of 60° in the room and an outside temperature of 30°, the proportion of loss of heat from wall-surface to loss of heat from window-surface may be approximately taken to be about 1 : 2·5.

CHAPTER XV.

PURITY OF WATER.

AFTER air, water is the first requirement for existence; impure water is as fertile a source of disease as impure air; and in judging of the hygienic qualities of water we must call to our aid the researches of the biologist as well as those of the chemist and physicist.

All available water comes from the rainfall; if rain falls on an impervious surface, it runs off in surface streams and rivers; if it falls on porous formations, it runs off in underground streams and rivers, and reappears in springs.

A cubic foot of water weighs 1000 ounces, or 62.5 lbs. An imperial gallon weighs 10 lbs. avoirdupois at 69° Fahrenheit. Fresh water in cooling becomes denser until the temperature reaches 39° Fahrenheit; after this it again expands until it reaches 32°, when it solidifies into ice. In the act of freezing, water expands considerably, and with sufficient force to burst iron pipes. This fact is well known; it is therefore astonishing that builders and architects continue year after year to leave water-pipes unprotected from frost.

There are few things which water does not dissolve to some extent. Its solvent powers are generally increased by rise of temperature, but there are some exceptions to this. Chloride of sodium (e.g. salt) is dissolved to the same extent whatever the temperature; sulphate of lime is less soluble in hot than in cold water.

Water absorbs gases in variable amounts—it dissolves ammonia and hydrochloric acid in large quantities; but it dissolves the gases of the atmosphere, oxygen, nitrogen, and carbonic acid only in small quantities, as will be seen from the following table :—

Quantity of different gases absorbed by 1 volume of water at 15° C (59° F) and 30 inches Barometric Pressure.

	Volume of gas dissolved by 1 vol. of water.
Ammonia	782·7000
Hydrochloric acid	457·8000
Sulphurous acid (SO_2)	43·5642
Carbonic acid (CO_2)	1·0000
Oxygen	0·0299
Nitrogen	0·0148

Water usually absorbs more gas the lower the temperature and the greater the pressure: when water is warmed it gives out the gases ; when it freezes the dissolved gases are *usually* liberated. This is a point of importance in the question of storage of water.

Rain, as it leaves the clouds, is pure, but in its passage through the air it absorbs certain gases, and carries with it particles of matter which may be floating in the air.

The gases it absorbs are oxygen, nitrogen, carbonic acid, a little ammonia, and nitric acid. The particles floating in the air are for the most part organic.

Rain water collected from a rocky surface sparsely occupied by population is the purest attainable water supply.

Near large towns other impurities are found in rain water. But notwithstanding this, such rain water is generally far purer than river water. The water found in rivers has either drained into the rivers from land, or, having fallen on porous strata, is given out from them in springs. For this reason all water in rivers or streams contains more or less of matters taken up from the soil. Thus rivers fed from a granite country will be comparatively pure, though frequently peaty. There is no evidence that a little peaty discolouration is injurious to health.

Rain which falls on cultivated land is liable to pollution

from the manure intended for the crops. When rain falls on a closely inhabited surface, or passes into a subsoil saturated with impurities, it will be contaminated.

The most agreeable waters are generally those containing nitrates and chlorides ; such waters should be avoided.

Pure water acts rapidly on unprotected iron pipes—cast iron is less affected than wrought iron—and a coating of asphalte or pitch will protect the iron, provided the coating be perfect. If pure rain water is allowed to come in contact with lead it will dissolve a part of the metal.

The action of water on lead appears to depend on the quantity of oxygen and carbonic acid. Where there is a large quantity of oxygen, the lead is rapidly oxidised, and the oxide of lead is to a certain extent soluble in pure water ; but if the water contains a sufficient quantity of carbonic acid to convert the oxide into carbonate of lead, which is only slightly soluble, the water will be comparatively safe from dangerous contamination.

A lead top and lead fittings in a pump of a well have been found to be injurious, from the evaporated water, which is pure, condensing on the lead, dissolving some of the metal, and dripping back into the well. Lead cisterns are subject to the same effect, and therefore, lead cisterns for storing water should be avoided.

Peaty matter in water forms a coating which protects the surface of the lead ; water which has been for some time in contact with unprotected iron pipes, and has been deprived of its free oxygen, is less liable to action on lead.

The conditions under which water acts on lead are so intricate that it is preferable, where any doubt exists, to use iron or earthenware pipes for its conveyance, and slate or earthenware cisterns for storing it.

Hard water never acts on lead ; it forms a protecting surface by the deposition mostly of sulphates, and carbonates of lime and magnesia.

The effect of hardness of water on health is a somewhat intricate question.

A comparison made on an average of five years between the death-rates of twelve towns furnished with soft water, as compared with twelve towns furnished with hard water, showed a marked preponderance in favour of the towns supplied with hard water as compared with those supplied with soft water, 10 degrees of hardness being the standard. But the sanitary condition of the several towns in other respects would require special examination before this evidence could be admitted as conclusive.

Of the conscripts taken in France, a larger number were found to be rejected on medical examination from soft water districts than those taken from hard water districts. On the other hand, Highlanders are a stalwart race, and the water they have is mostly soft water.

Hard water is alleged to produce in some individuals certain diseases, such as stone ; but, on the other hand, in valleys in mountainous districts, where the water is practically pure rain water, the inhabitants suffer from goître.

The economical advantages for domestic use of soft water are undoubted, arising especially from the saving of soap. Each degree of hardness destroys $2\frac{1}{2}$ ounces of soap in each 100 gallons of water ; therefore, soft water is commercially of more value than hard water, in the proportion of 5 ounces of soap to each 200 gallons for each degree of hardness. For many manufacturing purposes soft water is preferable. Thus, dyeing requires soft water,—on the other hand, ale-brewing requires hard water.

Water of 10 degrees of hardness would probably satisfy the general requirements of a town supply.

The terms ' hard ' and ' soft ' refer to the soap-destroying power of a water. Soap is an alkaline stearate. The addition to it of lime and magnesia decompose it, forming a calcic or magnesic stearate. Hence the reason why it is difficult to obtain 'a lather' with hard water (i. e. a water containing lime and magnesian salts), because a certain quantity of the soap is required to decompose the calcic and magnesian salts before a lather can be obtained, whilst, conversely, a lather is at once formed with a soft water, because of the absence of calcic and magnesic salts.

Two kinds of hardness are usually described—

Temporary hardness is due to calcic or magnesic carbonates. These salts are almost insoluble in pure water, but are freely soluble in water containing carbonic acid. On boiling, the carbonic acid is expelled, and the carbonates are precipitated. Hence temporary hardness is that hardness which may be got rid of by boiling the water.

Permanent hardness is chiefly due to the presence of calcic and magnesic sulphates, and is not got rid of by boiling.

In expressing the hardness of a water in degrees, it is to be understood that every degree, theoretically, represents one grain of calcic carbonate, or its equivalent in soap-destroying power, in one gallon of water.

Hence, the universally-employed test for hardness is the soap test, originally suggested by Dr. Clarke. This test consists in employing a solution of soap of known strength, and ascertaining how much of this solution is required to form a lather which will last a certain time.

For the purpose of this test, in the first place, 16 grains of neutral chloride of calcium (prepared by solution of carbonate of lime in hydrochloric acid, and repeated evaporation to dryness, until all the excess of acid is driven off) are dissolved in a gallon of distilled water, which is then said to be of 16 degrees of hardness, a solution of soap of a given degree of strength is then formed. The former of these liquids is used to graduate the latter. The degree of hardness of the water to be tested is estimated from the number of measures of soap solution required to form a permanent lather.

Chalk waters are notoriously hard waters. They are partially softened by boiling, and deposit a large amount of fur.

In selecting a water for use it is necessary to determine—

1. The initial hardness.

2. The permanent or irremoveable hardness.

3. The removeable or temporary hardness.

Chalk water may be softened to a considerable extent by another process invented by Dr. Clarke.

To explain this process it is necessary to consider the composition of chalk.

A pound of chalk consists of—lime 9 oz., carbonic acid (CO_2) 7 oz.

The 9 oz. of lime may be obtained separately from CO_2 by burning, e.g. driving off the 7 oz. of carbonic acid by heat. When so separated the 9 oz. of lime may be dissolved in not less than 40 gallons of water, and what is called *lime-water* obtained.

The 16 oz. of chalk would probably require 5000 gallons of water to dissolve them—so sparingly soluble is chalk.

But if 7 oz. more of carbonic acid, in addition to the 7 oz. of carbonic acid in the chalk, be added to the 16 oz. of chalk in water, then it becomes readily soluble, and bicarbonate of lime is formed. If the proportion of such a solution were 16 oz. of chalk and 7 oz. of carbonic acid to 400 gallons of water, then the water would resemble well water from the chalk strata.

Now if a solution of lime-water, e. g. water containing 9 oz. of lime, be united to a solution of water containing 16 oz. of chalk and 7 oz. of carbonic acid, then the solutions will so act on each other as to form 2 lbs. of chalk. Thus :—

Bicarbonate of lime in 400 gallons. { 1 lb. chalk = 16 oz. { Lime 9 oz. { Carbonic acid 7 oz. } = 16 oz. } with { Carbonic acid 7 oz. }

Burnt lime in 40 gallons of water 9 oz. { ... ... = 16 oz. } = 16 oz. } = 2 lbs. of Chalk.

In point of fact, only $\frac{10}{11}$ of the chalk would be separated—

thus the water of $17\frac{1}{2}$ degrees of hardness would remain of $1\frac{1}{2}$ degrees of hardness.

In practice, however, chalk water contains other hardening matters besides chalk; and consequently the softening can rarely be effected below $2\frac{1}{2}$ or 3 degrees. But the process also appears to remove mechanically micro-organisms as well as a great portion of any organic matter in the water.

Where water has to be stored, or to remain exposed, as in reservoirs, for any length of time, there is on the one hand the great liability of soft water to absorb noxious gases, and on the other the tendency with hard water to foster vegetation of various kinds. Waters from the new red sandstone are worse even than the chalk waters in this latter respect. On the other hand, the storage of water assists its purification. Impurities fall to the bottom: hence in taking water from a reservoir it should be drawn from a level not far below the surface. When filtered water is stored it is best to employ covered cisterns, because although light is unfavourable to the growth of micro-organisms it is favourable to vegetation, which would furnish the water with decaying organic matter.

The general principles of water-supply may be stated briefly as follows:—

1. To select the purest available source after careful analysis.
2. To filter the water, if necessary, in order to free it from suspended matter and from dissolved organic matter and from micro-organisms.
3. To store it in covered tanks, and to raise it a sufficient height for distribution by gravitation.

Water may be obtained from—

mountain ranges, which act as condensers;
rivers and streams;
natural springs;
wells artificially formed;
impounding reservoirs;
a combination of two or more of the sources named.

And water may be conveyed for distribution by means of—

    open conduits (before filtration);

    covered conduits, always, after filtration;

    cast-iron pipes under pressure,—where a district is to be supplied with water.

Water containing much suspended matter is generally to be avoided, but frequently the suspended matter may be removed by filtration or simple straining.

The mere existence of organic matter either in solution or in suspension is no proof of impurity. If water contains the fresh juices of inoffensive plants, it would not be unwholesome. But when these juices putrefy, or when water contains organic matter, and especially animal matter, ready to putrefy, it should be avoided. Processes of fermentation and putrefaction are due to living organisms, and living organisms appear to be connected with some form of zymotic disease; for the evidence is very strong that Asiatic cholera and typhoid fever may be communicated in drinking water. Therefore the chief interest which attaches to the presence of organic matter in water is that it may serve as an indication of the possible presence in the water of living organisms endowed with virulent properties. The recent advances in bacteriological study have opened out the possibility of solving many problems connected with water supply which had hitherto presented great difficulties. But bacteriological investigation requires special appliances, skill, care, and time. And in the case of expeditions into new countries, or in the case of an army on the march, a rapid diagnosis of the quality of water may be imperative.

A few remarks on simple facts connected with examination of water may therefore be useful.

The quality of a water can only be judged of from its constituents as a whole. The values of these constituents are not their values absolutely, but their values relatively, and in connection with the history of the water. Fifty grains of salt per gallon may mean nothing; five grains per gallon may mean danger.

Care must be taken that the specimen of water to be examined is a normal sample of the river or spring to be tested, that the vessel it is placed in is cleaned and rinsed out with some of the water to be tested. The water should be collected from a spot where it is not subject to artificial disturbance, and in filling a bottle its mouth should be immersed, if possible, from 1 to 2 feet below the surface.

The bottle should be kept stoppered, and allowed to stand for a day or two, exposed to light but not to evaporation, to see whether vegetation or putrefaction is developed. It should be in a temperature suited to vegetation.

If animalculæ appear, they are an indication of nitrogenous matter, and are one proof of the presence of substances capable of putrefaction; a microscope is necessary to detect the smaller forms of life.

The affinity for oxygen possessed by a sample of water affords a measure of organic impurity : and a solution of permanganate of potassium enables the amount to be easily ascertained. In the presence of oxidisable matter this substance freely parts with its oxygen until the permanganate has been reduced to hydrated manganese dioxide. This is shown by the pink colour of the solution being altered to brown. But as temperature and variation of acidity affect the operation, if anything beyond a rough test is required it must be conducted by a skilled operator. In a general way it may be said that pure water will not consume more than 0·05 parts of oxygen per 100,000 parts of water in 15 minutes, at a temperature of 80° Fahrenheit.

Clearness and absence of colour in water is no certain indication of its being wholesome. The worst waters may be clear, bright, and colourless. Colour as a companion to chemical analysis may give to an experienced observer a clue

P

to the kind and quantity of organic matter present. Its colour is best judged of through a tube 2 feet long, 2 inches in diameter.

If a water exhibits a bluish tint, or, say, appears nearly colourless in the 2-foot tube; if it, moreover, uses up very little oxygen after standing for three hours in contact with permanganate, the freedom of that water from organic impurity may be relied upon as certain.

If a water exhibits but little colour, or at most a slightly yellow, or a greenish-yellow tint, but if the oxygen it uses up is found to be large, such a water as a potable water is suspicious.

If a water exhibits in the 2-foot tube a decided peaty tint; if by experiment it is found to need a large quantity of oxygen after standing for three hours; knowing that peaty matter acts as a reducing agent on permanganate, the quantity of oxygen required, although far in excess of what was used in the former case where there was an absence of colour in the water, is not to be regarded with the same suspicion, peaty matter not being injurious to health.

Water that will not bear the test of standing will, in most cases, be rejected at once. If no other water can be obtained, it ought to be used before putrefaction has set in, but this is a great risk; the next best method is to wait until after putrefaction has terminated.

An indication of the nature of organic matter in water may be obtained by evaporating a small quantity of the water, weighing the residue, and then burning it to drive off the volatile matter, and comparing its weight before and after burning. If the residue blackens in burning, it indicates animal organic matter.

In considering whether organic matter comes from animals or vegetables, the presence of common salt may in many cases, with proper precautions, be found to be a nearly certain guide.

We consume not less than 100 grains of salt a day. This salt is thrown off daily. From all animals there is a large outflow of salt. It is the constant accompaniment of the animal living, or decomposing after death.

If much salt is found in water containing organic matter, nitric acid will generally be found, and if not nitric acid, then animal matter unoxidised.

In the case of dead animals organic matter is destroyed or retained in the soil ; phosphates and other organic substances are also retained. Salt is removed by water. Sewage contains chlorides, and the amount of salt is the most certain method of ascertaining the quantity of sewage which is or has been present—assuming that there are no disturbing causes.

Of course this test must be used with caution. Near the sea the spray is driven many miles inland. In many districts, where deposits of salt exist, wells and springs are saline. Chlorides are also given out from manufactures.

Water containing chlorides to a great extent ought not to be used without careful examination as to the source—three grains per gallon of common salt coexisting with an excess of nitrates is a cause of suspicion.

The average of the chlorides in the water of a district having been obtained, any increase of common salt above the average in a well situated in a camp or city, or near habitations, is an almost sure test of impure drainage.

Although caution must be exercised in drawing conclusions from the presence of chlorides, their absence is conclusive against the presence of decomposed animal matter or excreta.

Nitrates are common in small quantities. The amount from atmospheric causes is very minute, but they are found in water from manured land, in gardens, in wells near houses, in towns, and in great abundance near churchyards.

If chlorides and nitrates are found together in water, it may be assumed that animal matter has existed, or does exist, in the water.

If there is much nitrous acid, it indicates recent organic matter or oxidation going on.

The examination thus suggested would show the presence or not of—

1. Organic matter decomposed or putrid.
2. Organic matter readily decomposed and probably ready to become putrid.
3. Organic matter slow to decompose, but still capable of becoming putrid.
4. From the nitrites recent organic matter.
5. From the nitrates old organic matter.
6. Vegetable organic matter.
7. Animal organic matter.

In deciding on the quality of water the relationship which each of its constituents bears to the others must be considered as well as the natural history and the physical condition of the water; and where time can be afforded a bacteriological examination is especially desirable. Any classification of waters founded on a single factor in the chemical analysis of water is to be accepted with great caution, however important that single factor may be.

The general line of examination of waters can only be indicated here.

The engineer in selecting a source of water-supply is so deeply concerned in the question as to what is good water, that these few practical hints on the subject may assist in the consideration of the subject.

But whilst much knowledge is derived from a water analysis, especially as to the waters which should be avoided, it is necessary in the case of waters reported by analysis to be good, to be very careful to keep up a continuous examination of their sources, and to see that these are not liable to contamination.

The sources of contamination of water arise at every step.

That which this country suffers from most largely is the contamination of rivers by sewage

With an increasing population over the districts which form the water-sheds of rivers, this is inevitable.

The fact that sewage or other contamination is poured into a river at a point high up on its banks is, however, no bar to the water of that river being used at some spot lower down for domestic purposes—the distance between the spot at which contamination has been poured in and that at which the water can be again used for domestic purposes, will depend upon the nature of the bed of the river, the rapidity of the current, and the temperature; the effect varying with the season, being much more rapid at a high than at a low temperature.

This cannot be better illustrated than by extracts from the Report of the Commission which reported on the pollution of the River Seine at Paris in 1875 and 1876.

Above Paris the Seine presents a satisfactory appearance.

The sewers of Paris discharge into the river black fœtid streams, covered with layers of greasy matter, which accumulates on the sides of the river. These streams transport particles of organic matter and débris, which are deposited as grey and black mud on the banks, or else form shoals. This mud is the seat of an active fermentation, throwing up innumerable bubbles of gas, which burst at the surface of the water; the bubbles attaining sometimes in hot weather a diameter of $1\frac{1}{2}$ metres (nearly 5 feet), dragging up the black mud with them to the surface. Fish and plants cannot exist. But as the river leaves these sources of pollution, it gradually improves.

From Epinay to Argenteuil the water is still of a deep colour, mud has disappeared, fish make their appearance. Below Bezons a most abundant vegetation clothes both banks, and large sheets of water-plants partly impede the course of the river. At Meulan all visible sign of pollution has disappeared, and the river is chemically pure.

The following table extracted from the Report of the Commissioners shows the chemical condition of the water:

Place.	Nitrogen not yet transformed into volatile ammoniacal salts, or organic nitrogen. Grammes per cubic metre or 1,000 litres.	Total nitrogen, including volatile ammoniacal salts. Grammes per cubic metre.	Dissolved oxygen in cubic centimetres per litre of water.
	Grammes.	*Grammes.*	*Cubic centimetres.*
Above Paris—			
Bridge of Asnières ...	0·85	1·5	5·34
In Paris—			
Clichy, below intercepting sewer ...	1·51	4·0	4·60
St. Denis, below intercepting sewer ...	7·27	7·0	1·01
Below Paris—			
Meulan	0·40	2·1	8·17
Vernon	0·40	1·4	10·40

The table explains the process which takes place.

The organic matters change into carbonic acid, water, ammonia, sulphuretted hydrogen, and different mineral substances. This change implies an absorption of oxygen from the gases dissolved in the water, and a production of mineral nitrogenous bodies.

As long as the water contains such matters susceptible of fermentation, it is unfit for use. When fermentation is accomplished, and the organic matter has passed into the state of mineral matters inoffensive in themselves, the water presents a disappearance of nitrogenous organic matter, replaced by nitrogenous mineral matter, by ammonia. The dissolved oxygen in the water is used up, but may be restored by movement, such as is caused by the flow of the stream, or more rapidly by agitation, as for instance in passing over a weir, and thus the water can be rendered fit for drinking.

It will be seen from this that although a river may have received sewage at one part of its course, there is some point below at which it will still become fit for use.

There are other and more occult sources of contamination.

All water proceeds from evaporation and rainfall. The rain which falls on impervious strata passes from the surface into rivers, whilst water which falls on pervious strata passes into the ground.

When this water, in its passage through pervious strata, meets with an impervious bed, it is arrested in its course ; and if the impervious bed dips down and forms a basin, then the water will remain in a subterranean reservoir accessible only by wells.

Where the impervious strata which underlie the pervious strata crop out at the surface, the water flows gradually down as an underground river, to pass out at the lowest point in springs.

An examination of the water-level of different districts seems to show that this general water-level forms an inclined plane, rising from the natural vent or outfall.

But the course of this underground river depends on the inclination of the water-bearing and underlying impervious strata, and not on the form of the surface.

Taking the water-levels of a line of wells at right angles to the outfall, be it in a river or in springs, it has been found that in the chalk the inclined plane of the surface of underground water rises at a rate of not less than 10 feet per mile [1]; in the boulder strata of Norfolk it appears to vary from 3 to 100 feet per mile ; and in certain of the tertiary beds it is about 5 feet per mile [2].

Of course this inclination varies with the head of water and the porosity or permeability to water of the material.

Mr. Roberts, of Liverpool, found by experiments in the new red sandstone of average coarseness of that district, that

[1] Clutterbuck.   [2] Baldwin Latham.

the following quantities of water passed per hour through a square foot of the sandstone 10½ inches in thickness :—

With a pressure of 10 lbs. to square inch, 4¼ gallons

　　　　　,,　　　　,,　　　　20　　,,　　　　,,　　　　7¾　　,,

　　　　　,,　　　　,,　　　　46　　,,　　　　,,　　　　19　　,,

which showed the increase to vary in a direct ratio with the pressure.

The level of the subterranean sheet of water, moreover, rises and falls with the quantity supplied by rainfall; thus, in wet seasons, the water-level approaches near to the surface, and in dry seasons it recedes. Mr. Baldwin Latham has shown the effect of this change in contaminating the water supply at Croydon. (Fig. 36.)

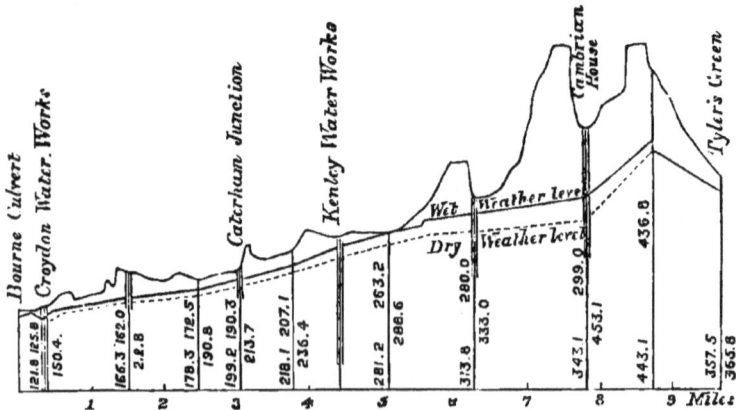

Fig. 36. Section of water-level in chalk near Croydon.

Croydon has been drained effectually, and the water-supply is obtained from springs at some distance off in the chalk.

In the dry seasons the supply was excellent; in wet seasons fever was found to prevail, which was attributed to the water. The cause of this was explained by the fact that in dry seasons the level of the water was far below any subsoil contamination; but that, in a wet season, the water-level of the underground river rose up to the level of cesspits which were not impervious, manure heaps, and other sources of contamination, and thus became contaminated with the contents.

There are, however, many causes of contamination of water which are much more obvious.

In a porous soil, the vicinity of a cesspit is a frequent cause of danger. Figs. 37 and 38 show how fluctuating may be the danger in such cases.

Fig. 37.

Fig. 38.

In one case (Fig. 37) the flow of underground water is from the cesspit to the well.

Whilst the ordinary level of underground water prevails, there would be nothing to indicate the danger; but upon an excess of rainfall occurring, by which the current of underground water was altered, then the well would become polluted, and sickness break out, to cease probably on a fresh change of conditions.

In the second case (Fig. 38) the flow of underground

water is from the well to the cesspit, and in this case there is
not the same imminent danger of pollution as in the former
instance.

The pollution of the water may occur in its transit from
the source of supply to the consumer. The London Water
Companies have sometimes had the water polluted by passing it through iron mains calked with hemp, when the hemp
has become a cause of injury to the water.

The question of water-supply would require a treatise
to itself; a few practical hints are all that can be given
here.

The quantity of water for the civil population cannot be
assumed at less than 15 gallons per head per day; for infantry
12 gallons per head per day, for cavalry 20 gallons per head
per day, in barracks.

An Artesian well (Fig. 39) is a well sunk through impervious strata, in which there is no water, into pervious strata

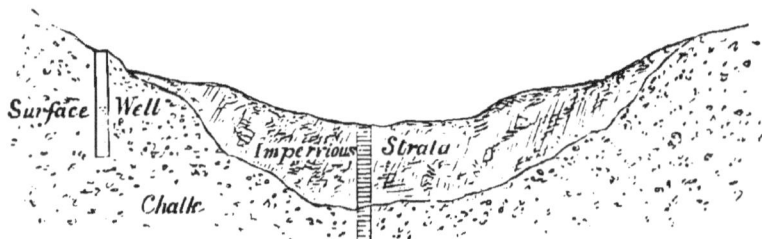

Fig. 39.   Artesian Well.

which derive their supply from a water-shed area at a distance. Thus, in London, wells are sunk through the London
clay to derive their supply from the chalk. The water in the
chalk comes from the water which falls on the chalk hills
surrounding the London basin. Deep well water is generally
very good and palatable water.

In porous soils water must generally be obtained from
surface wells.

A spring is the lowest point or lip in the stratification of an underground reservoir of water. A well sunk in such strata will most probably furnish, besides the volume of the spring, an additional supply of water.

Well water will vary in purity according to the nature and amount of the soluble matters contained in the ground from which the well derives its supply.

Shallow wells are always liable to pollution from vegetable matter, or animal matter, in or on the surface soil.

In deep wells differences have been observed in the purity of well water, according as the water was taken from the surface by dipping, or was pumped from the bottom of the well. In one case, where the depth of water was over 50 feet, the surface water was found to be yellowish in colour, much harder, and more contaminated with organic impurities than that from the bottom of the well, as if a layer of lighter water from the surface drainage floated on the spring water below. In another case the solids per gallon amounted to 66·23 grains in the bottom water of a well and to 3 grains only in the surface water.

Such differences are often constant. Exhaustive pumping in wells may be injurious to the water ; especially in wells in the vicinity of the sea, when the brackish water usually kept back by the land water may be drawn in.

Norton's tubes are useful for ascertaining the quantity of subsoil water, and for drawing the supply from a certain depth. They are easily applied ; but the supply from each has necessarily a limit.

In camps it may frequently be necessary to resort to a surface supply of water.

The following are a few temporary expedients for camp purposes of water supply.

Where the water was near the surface, one system pursued by the Turks in the Crimean war was to select a site near their camp where the surface was clean, and dig a hole and place

in it a barrel (Fig. 40), in the bottom of which holes had been
bored, so as to ensure that water from the deepest part of
the hole should alone come into the barrel. If they wanted
further filtration, they got a smaller barrel, and bored holes in
the sides as high as was desired, and they then placed the
smaller barrel in the larger one, ramming sand in below and
all round so as to bring its top to
the level of the top of the first barrel,
and thus forming an upward filter of
sand through which the water passed.

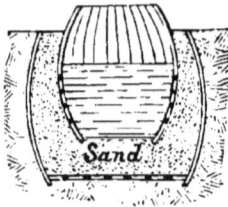

Turkish Camp Well.
Fig. 40.

The Russian plan, where brushwood was obtainable, was to make a
large gabion 4 or 5 feet in diameter,
some 10 or 12 feet long, and then to
dig the well and drop this gabion
into it. This formed a temporary casing, which answered its
immediate object.

For wells of any degree of permanence, care is necessary to
ensure that the surface water for a certain distance round the
well should not pass back into it, for the surface near a well
is always liable to pollution.

In deep wells, i.e. wells in which water of the requisite
purity is only found at a considerable depth, the surface soil
water should be cut off from the deep water by carefully
casing the well and puddling behind the casing above the
water level, so as to prevent surface water from trickling
down the sides.

Surface wells in porous strata should be lined with puddle
behind the steyning to the full depth, or in wells beyond 12
to 15 feet in depth, then to that depth at least, so as to
ensure complete filtration of the surface water which passes
into the well. The surface round the well should also be
puddled, sloped away from the well, and paved, so as to
prevent the dirty water which is thrown down near the well
from finding its way back again. (Fig. 41.)

The distance to which this preparation of the surface should be carried round the well depends on the depth of the well; with a deep well, in which there would be a considerable depth of soil through which the surface water would have to pass before it reached the well, an extensive surface covering would not be necessary; whilst with a shallow well,

by extending the area of surface puddling, the amount of filtration to which the surface water would be subjected before it reached the well may be materially extended.

The tube lining of the well should be carried up to at least two feet above the surface, so as to prevent surface impurities from falling in. The well should

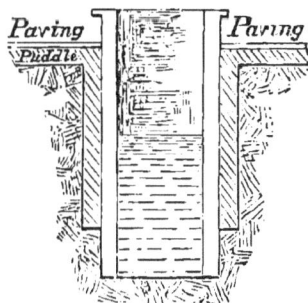

Fig. 41. Shallow Well.

be covered, as a protection against dead leaves, &c.

For keeping well-water clean, it is preferable that it should be drawn up by a pump. In the use of buckets impurities frequently fall in. If not a pump, then an iron chain and bucket, because they can be kept cleaner than rope and wood.

In some cases there is advantage in the aeration which the water obtains by being exposed to the air. In such cases, closing the well and drawing water by pump has rendered the water undrinkable.

Where a river flows through a valley over porous substrata, sinking a well or wells in the strata within the influence of the river filtration is a cheap and ready method of obtaining river water naturally filtered. The tops of wells so situated must be carried above the level of extreme floods.

If a single well on a river bank does not produce sufficient water, or if the subsoil is clay, impervious to water, trenches may be excavated parallel to the river or stream, in which trenches perforated earthenware pipes may be laid, leading to a well or wells. The trenches above such pipes should be filled in with fine assorted gravel, charcoal, and sand, so as to form a filtering medium within the reach of the dry weather flow of such stream or river. These trenches should not be less than 6 feet deep to the top of the pipes.

At Florence part of the water-supply for the town is derived from channels nearly a mile long, carried by the side of and below the level of the Arno.

Natural springs may be utilised by storing the water in a reservoir which will contain the flow of one entire day, or of a longer period.

All reservoirs and tanks should be constructed to prevent infiltration of water from the soil, the sides and bottom should be lined with puddle, faced with brick or masonry walls, and coated with cement. Reservoirs may be covered in to protect the water from contamination. In case of peaty waters aeration is beneficial; and in such cases it may be advisable to draw off the upper film of water for use.

In positions where there is a difficulty as to wells, springs, or streams, dependence must be placed on rainfall. Where water is to be collected from rainfall, great care is requisite in order to keep perfectly clean the surfaces on which it falls, as well as the conduits to the reservoirs and tanks.

When rain water is collected from roofs, the first washings of the roof should not be collected.

Rain water is not agreeable to the taste until it has been stored in the ground, where it absorbs carbonic acid gas.

The rain water from roofs should be first collected in a settling tank, of a sufficient area to allow the heavier sediment to be deposited; it should then be passed through a gravel or coarse sand strainer into the tank or cistern for use. It is customary in parts of Italy to provide three receptacles. The rain water is received and allowed to settle in the first, it is strained off into the second, where it is retained if possible for a year, and then passed into the third for use. The settling tank and the strainer should be cleaned out frequently.

In the case of porous strata of clean sand or gravel, where sufficient ground can be set apart, and where military or cheap labour is available, water may be stored by excavating an area, of a size calculated on the minimum rainfall, to a depth of 5 or 6 feet, and paving the bottom with tiles, or impervious material such as puddle, cement, or asphalte, sloped down towards one corner, and lining the sides with puddle, and then filling in the whole again with the porous material of the ground. A pump could be fixed in the lowest corner. The surface should be covered with grass, and kept free from impurities. By these means a valuable reservoir could be formed, which would keep the water cool. This idea was originally described by Bernard Palissy.

All water, before being stored in tanks, from which it is to be pumped direct for use, should be passed through a filter; every rain-water storage tank should have its filter.

Recent investigations have considerably altered the formerly accepted views on filtration.

A filter was formerly regarded simply as a strainer. In this capacity the most perfect filter is that of Pasteur— formed of unglazed biscuit porcelain, which will even remove bacteria from the water; but it requires that the water be supplied under pressure. The surface requires frequent cleansing: and this form of filter is only applicable on a small scale. The efficiency of the strainer depends upon the surfaces presented by the material.

For filters on a large scale, fine sand is the simplest and best material. These surfaces serve more or less to attract solid matter brought within range of their attraction. It has been estimated that a cubic yard of sand would contain an area of surface in the particles of about 2,500 square yards.

A sand filtering-bed should have a thickness of about 2 feet of sand at the top, under which should be placed four layers of gravel, each 6 inches thick, the first of the size of shot, the next of peas, the third of beans, the fourth the size of walnuts ; the rate of filtration should not exceed from one and a half to two gallons per superficial foot per hour.

The water often passes rapidly at first, and the filtration through the sand-filter is therefore sometimes less effectual when first constructed ; when the surface becomes clogged with a film of slime, the rate at which the water passes decreases. This film of slime acts beneficially to prevent bacteria passing through.

When however the rate of filtration is diminished below 1.5 gallons per superficial foot per hour, it becomes necessary to replace the upper film of sand by a fresh layer of sand.

When the whole stratum of sand has become filled up with organic matter, the filtering bed should be renewed. The duration of a filtering bed in fairly constant use may be assumed as a rule not to exceed two years.

If the bed were left dry at frequent intervals, a new condition would arise and the filter would cease to be a mere strainer, and become an apparatus for enabling bacteria to act as agents for converting organic impurity into harmless mineral matter.

Filters for domestic purposes were experimented on by Dr. Parkes. He showed that a sand filter, which has considerable purifying power when first used, clogs rapidly. The sediment stops mainly on the surface, and thus diminishes its filtering power ; this can be restored by scraping off a small film of the top surface of the sand.

Magnetic oxide of iron may be usefully added to a sand filter where there is much organic matter. It increases the oxidising power of the filter, and renders it more effective for destroying organic matter.

Sponges are useful with sedimentary waters ; they arrest more than their own weight of solid matter, and are very easily cleansed.

One of the best filtering mediums is animal charcoal. The charcoal, in consequence of the very large surface arising from its porosity, undoubtedly has great power as a mechanical filter. It has been contended by some chemists that the organic matter which passes into it becomes oxidised. On this point, however, the evidence is not conclusive.

Spongy iron appears to be an active agent, not only in removing organic matter from water, but in reducing its hardness and altering its character when water is filtered through the spongy material.

Filters retain the impurities which are present in the water, and at first the outflowing stream is materially purified; but the retention of the impurities in the filter becomes very soon a source of danger of itself.

The pores become filled with the dead bodies of the organisms which were living in the water, and the water passing through may become after a time more dangerously polluted than it was before it entered the filter.

Filters must, therefore, be frequently and thoroughly cleansed by washing the materials of which they are composed, and by exposing these materials to the air, so as to cause any remaining impurities to be oxidised.

A filter with sponge to arrest the sediment, so arranged that the sponge can be frequently cleansed, and a porous filtering medium so arranged as to have a free exposure to the air when water is not actually passing through, would be the best form of domestic filter.

After filtration water should be kept in covered receptacles for use.

The storage and distribution of water is the next consideration.

The best position for keeping water is in tanks underground, assuming always that it is not in proximity with sources of impurity. In the ground it will retain an even temperature, and will take up carbonic acid gas, which makes it pleasant to the palate. When stored in cisterns above the ground-level, where its temperature is necessarily variable, much danger may ensue from its power of absorbing gases.

For instance, on a warm day, the water becoming warm, gives out the oxygen or other gases it contains; at night, when it cools down, its capacity for absorbing gases is much increased; and if there are impure gases near, such as sewer gas, or ammonia from dung heaps or manure heaps, it will absorb them in large quantities, and thus become dangerously polluted.

The system of water-supply which minimises this source of danger is to be preferred. Thus well-water was observed to have a maximum temperature of 51·5°, and a minimum temperature of 50·5°, or a mean of 51° during the year in the same locality. The water supplied from mains had a maximum temperature during the year of 64·8°, a minimum temperature of 39·7°, and a mean temperature of 52·25°; and the water in cisterns above ground varied from 71·5° maximum to 33·60° minimum, with a mean of 52·55°. Thus whilst the range of temperature of the well-water was only to 1°, the range of the water in the cistern was nearly 38°.

There is danger from the use of cisterns unless frequently cleaned, for the cistern becomes a place of deposit for any impurities which there may be in the water; especially if the cistern is left unused for a time; and thus, like the filter when not cleansed, cisterns may render the fresh water which comes into them impure.

Moreover, the supply of water to cisterns is necessarily intermittent, and when cisterns have been filled, and the pipes are emptied of water, cases have occurred in which impure gases have been drawn into the pipes, and have been taken up by the water when it was again turned on into the pipes.

A case occurred in which a town council, in order to economise the laying of pipes, carried them in the sewers.

A severe epidemic of typhoid fever occurred. This was ac-
counted for by the assumption that sewer gas was drawn
into the pipes, either through the joints or by diffusion
through the pores of the metal.

On these grounds the constant service, in which the pipes
are always filled under pressure, is preferable to the intermittent supply.

As water will readily absorb foul gases, and may become
poisonous, the possibility of contamination from foul air should
be prevented. Hence any overflow or waste-water pipe from
a service reservoir or tank should deliver the water at an open
end into a channel, and be thence conducted into the covered
sewer, or drain, so as to prevent gases from the sewer rising
back through such overflow or waste-water pipe to the water
in the reservoir or tank.

Water conduits should be laid at such depth and be so
covered with earth as to prevent the water becoming heated
unduly by the rays of the sun, or being injuriously affected by
frost in winter. This depth may be considered to be not less
than three feet.

All covered reservoirs and tanks should be ventilated,
and so situated as to be easily emptied, inspected, and
cleaned.

All supply-pipes should be arranged in such manner as to
allow of easy inspection and subsequent repairs. Stop-taps
should be placed betwixt the water main and the building in
all cases, so as to allow of the isolation of any line of service-
pipe for repairs.

All house service-tanks and service-pipes should be fixed in
such manner as to be carefully protected from frost, and so
that the best rooms shall not be flooded on the occurrence of
leaks or overflows. When protected by wooden casings, the
front of the casing should be made to open easily.

The distribution of water in camps should invariably be
placed under inspection, and under the charge of some person

who should be responsible for not allowing waste or pollution. With this object the general dipping of buckets into a well should be prevented.

If a pump is not fixed, there should be one man made responsible for raising the water out of the well. He should pour it into a box or reservoir in front provided with taps, out of which all who come for water should draw without confusion or fouling of the well with dirty buckets.

The surface of the ground near any place where water is obtained should, if possible, be paved, or else it will soon become a mass of mud. This is especially desirable with horse-troughs.

In making troughs for watering horses care should be taken to supply each trough independently of the neighbouring troughs. It has been often the practice to place the troughs in a line, and let the overflow of the first fill the second, and so on; but the water of a trough becomes fouled by the horses drinking in it, and consequently the overflow from the upper troughs conveys nothing but polluted water into the lower ones.

No vessel to convey water for drinking purposes for human beings should under any circumstances be allowed to be dipped into a trough used for watering horses.

CHAPTER XVI.

REMOVAL OF REFUSE.

A THEORETICALLY perfect system of refuse removal would be one where a large volume of rapidly flowing water received the whole refuse, and carried it away, before it had time to decompose, to a large river not used for drinking purposes, and thence to the sea ; but this is generally unattainable, and it would leave the waste matter unutilised.

Refuse from dwellings falls under the heads of : —

1. Ashes.
2. Dry kitchen refuse.
3. Stable manure.
4. Solid or liquid ejections.
5. Fouled water from washing and cooking.
6. Rain water and washings of passages, stables, yards, and pavements.

The first three of these must be removed and disposed of by hand labour. The fourth may be treated on the dry earth system ; but in houses, and especially in houses without gardens, is preferably removed by water carriage. The two last in any case must be termed sewage.

To arrive at a clear conception of what the problem is, it must be remembered first, that even where a water-carriage system prevails in any town, some plan of removing house

refuse, ashes, and dust—some 'dry system' of collection—
must also be in force, and secondly, that an ordinary household of, say, six persons, enjoying a proper water supply, will
consume for all purposes from 15 to 20 gallons per head
per diem, or a total daily quantity of from 90 to 120 gallons.
which must necessarily be fouled in its use, and pass away
from the house, laden with dangerous impurities. If earth-
closets are used instead of water-closets, a certain quantity
of water, perhaps as much as four or five gallons per head
may be saved, but there will be still from 10 to 15 gallons
of water per occupant fouled with grease, soapsuds, vegetable
refuse, and other materials of a highly putrescible kind,
which must necessarily pass away and be disposed of at the
outfall.

Thus some system for the removal of foul water is absolutely
necessary in every town, and the Rivers Pollution Commissioners, in their First Report, after a careful comparison
of thirty-one towns, in fifteen of which the midden system
prevailed, while sixteen were water-closet towns, came to the
conclusion that it mattered little, as regards the degree of
pollution in the sewer water, whether a system of interception
of the excreta was practised or not.

Hence, with the dry conservancy system, there must be
the removal of sewer water.

With the water-carriage system the removal of ashes, refuse,
and stable-yard manure has still to be provided for; but the
removal of ashes and stable-yard manure is a much simpler
matter than the removal of excreta.

The great principle to be observed in removing the solid
refuse is that every decomposable substance should be taken
away at once.

This is especially necessary in warm climates.

The principle may, however, be applied in various ways
to suit local convenience. In open situations, exposed to
cool winds, there is less danger of injury to health from

decomposing matters than there would be in hot, moist, or close positions. In towns in warm climates all refuse should be removed daily from both the house and its enclosures.

In the country, generally, there is less risk of injury than in the close parts of towns. These considerations show that the same stringency is not necessarily required everywhere. Position by itself affords a certain degree of protection from nuisance. The amount of decomposing matter usually produced is also another point to be considered. A small daily product is not, of course, so injurious as a large product. Even the manner of accumulating decomposing substances influences their effect on health. There is less risk from a dung-heap if protected from wet, and if placed to the leeward, than to the windward of an inhabited building.

The receptacles in which refuse is temporarily placed should never be below the level of the ground.

If a deep pit is dug in the ground, into which the refuse is thrown in the intervals between times of removal, rain and surface water will mix with the refuse and hasten its decomposition, and generally the lowest part of the filth will not be removed, but will be left to ferment or putrefy, and the presence of this fermenting mass will tend to promote the decomposition of the fresh refuse added to it.

Even the daily removal of refuse entails the necessity of places for the deposit of refuse. In the selection of these the following conditions should be attended to :—

1. That the places of deposit be sufficiently removed from inhabited buildings to prevent any smell being perceived by the occupants.

2. That the places of deposit be above the level of the ground—never sunk in the ground. The floor of an ash-pit, or dung-pit, should be at least three inches above the surface-level.

3. That they be lined with non-porous material. That the floor be paved with square setts, or flagged, and drained.

4. That they be protected from rain and sun, but open to the air.

5. That a space be paved in front, so as to prevent the traffic, which takes place in depositing the refuse or in removing it, from cutting up and polluting the surface.

A moveable iron box is a good receptacle for kitchen waste. For stable-yard refuse, the moveable wire-work grating in use in London is safer than any form of dung-pit.

In camps the disposal of manure and offal requires great care ; burning, unless carefully done, may give rise to much nuisance ; burying at a safe distance is generally the most advisable method.

The disposal of the solid or liquid ejections, which constitute what is generally termed sewage, must next be considered ; and before entering upon the general question, a few of the rules which govern temporary emergencies in camps may be mentioned.

A camp unprovided with latrines is always in a state of danger from epidemic disease.

One of the most frequent causes of an unhealthy condition of the air of a camp is, either neglecting to provide latrines, so that the ground outside the camp becomes covered with filth ; or constructing the latrines too shallow, and exposing too large a surface to rain, sun, and air. Latrines should be so managed that no smell from them should ever reach the men's tents ; to ensure this, very simple precautions only are required.

1. The latrines should be placed to leeward with respect to prevailing winds, and at as great a distance from the tents as is compatible with convenience.

2. They should be dug narrow and deep, and their contents covered over every evening with at least a foot of fresh earth. A certain bulk and thickness of earth are required to absorb the putrescent gas, otherwise it will disperse itself and pollute the air to a considerable distance round.

3. When the latrine is filled to within two feet six inches or three feet of the surface, earth should be thrown into it, and heaped over it like a grave to mark its site.

4. Great care should be taken not to place latrines near existing wells, nor to dig wells near where latrines have been placed. The necessity of these precautions to prevent wells becoming polluted is obvious.

Screens made out of any available material are, of course, required for latrines.

This arrangement applies to a temporary camp, and is only admissible under such conditions.

A deep trench saves labour, and places the refuse in the most immediately safe position; but a buried mass of refuse will take a long time to decay; it should not be disturbed, and will taint the adjacent soil for a long time. This is of less consequence in a merely temporary encampment, but it might entail serious evils in localities continuously inhabited.

The following plan of trench has been adopted as a more permanent arrangement in Indian villages, with the object of checking the frightful evil of surface pollution of the whole country, from the people habitually fouling the fields, roads, streets, and watercourses.

Long trenches are dug, at about one foot or less in depth, on a spot set apart, about 200 or 300 yards from dwellings. Matting screens are placed round for decency. Each day the trench which has received the excreta of the preceding day is filled up, the excreta being covered with fresh earth obtained by digging a new trench adjoining, which when it has been used is treated in the same manner. Thus the trenches are gradually extended, until sufficient ground has been utilised, when they are ploughed up and the site used for cultivation.

The Indian plough does not penetrate more than eight inches, consequently if the trench is too deep, the lower stratum is left unmixed with earth, forming a permanent cesspool, and becomes a source of future trouble.

It is to be observed, however, that in the wet season these trenches cannot be used; and in sandy soil they do not answer.

This system, although it is preferable to what formerly prevailed, viz. the surface defilement of the ground all round villages, and of the adjacent water courses, is fraught with danger, unless subsequent cultivation of the site be strictly enforced, because it would otherwise retain large and increasing masses of putrefying matter in the soil, in a condition somewhat unfavourable to rapid absorption.

These arrangements are applicable only to very rough life or very poor communities.

In permanent camps and barracks, and in villages where the houses or cottages are close together, and where therefore sufficient garden ground cannot be allotted to each to allow of the refuse being absorbed, which is especially the case with large Indian villages, some one of the methods adopted for town conservancy hereinafter described would usually be found applicable.

Midden, Pail, and Dry Earth systems.

Moveable apparatus, such as middens, pails, or dry earth closets, all involve the retention of the excreta for some period of greater or less duration, as contradistinguished from the immediate removal which takes place when water carriage is used. Moreover the weight of the tub, pail, or receptacle in which the excreta are removed has to be added to the weight of the refuse to be carried away, whilst in the water-carriage system the water itself is the vehicle of transport.

There are numerous towns, and innumerable villages and cottages, where the midden closet or privy with a fixed receptacle still prevails.

In its old form the receptacle consists of a pit with sides of porous materials, permitting free soakage of filth into the surrounding soil, capable of maintaining the entire dejections from one or more houses for months or years, uncovered and open to wind and rain and sun, undrained and difficult of access for cleansing.

The first improvement was to provide a cover to keep out the wet; the next, to make the sides and bottom of the pit impervious, and to provide a drain to carry off the excess of liquid: this was objected to because the effluent from such a pit is necessarily putrid, and becomes a source of danger in any drain or ditch into which it passes. The next important improvement was to reduce the size to a mere space behind the seat, formed of some non-porous material, such as glazed brick, and arranged for easy cleaning from the back. The small size prevents the retention in the vicinity of the dwelling of a large mass of putrescent matter. Privies of this class are, however, inadmissible except in detached positions away from dwellings, as for instance at the end of a cottage garden in the country; but where used, the floor of the privy should be raised, so that the seat may be high enough to admit of the floor of the receptacle being on or a little above the level of the ground, with a slightly raised edge to prevent any liquid flowing over. The receptacle received dry refuse and ashes, which helped to deodorise the contents and soak up the liquid; and had a cover to prevent rain falling into it, but so placed as to allow of the circulation of air under the cover.

The next transition from this was to a moveable receptacle. Of this type the simplest arrangement is a box or bucket placed under the seat, which is taken out, the contents emptied into the scavenger's cart, and the box cleansed and replaced.

A further alteration was the separation of the solid and liquid excreta; a system which has not however attained to much practical application.

Excrement Pail full, with lid on, to return to Works.

Excrement Pail empty, ready for use.

Figs. 42, 43.   The Rochdale Pail.

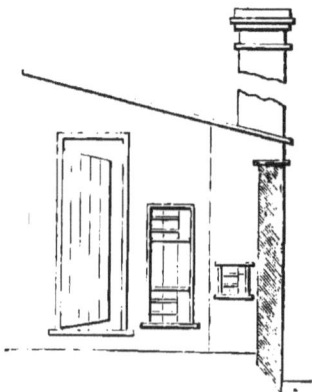

Elevation in Yard.

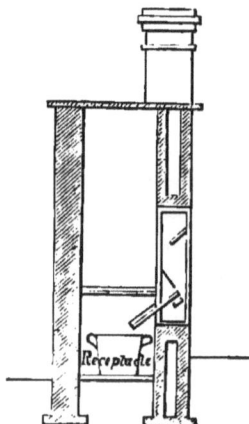

Sectional Elevation.

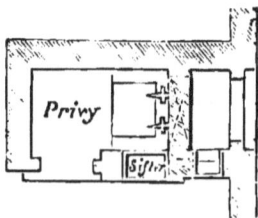

Plan above Privy Seat.

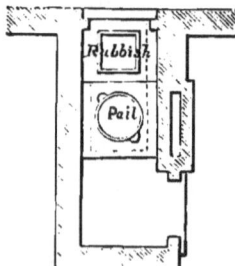

Plan of Privy Floor.

Figs. 44, 45, 46, 47 .   Closet with pail and refuse box.

The difficulty of cleaning the angles of the boxes led to the adoption of oval or round pails. (Figs. 42 and 43.) The pail is placed under the seat, and removed at stated intervals, or when full; some form of deodorising material is thrown in daily. In the north of England the arrangement generally is that the ashes shall be passed through a shoot, on which they are sifted—the finer fall into the pail to deodorise it, the coarser pass into a box, whence they can be taken to be again burned—whilst a separate shoot is provided for kitchen refuse, which falls into another pail adjacent. (Figs. 44, 45, 46, and 47.)

The general arrangement is for the floor of the privies to be somewhat raised, so that the space underneath the seat in which the tub is placed may be at or somewhat above the level of the ground, so as to be easily cleaned.

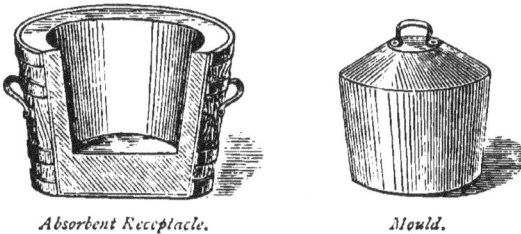

Absorbent Receptacle. *Mould.*

Figs. 48, 49.

Lined Pail used for Goux System.

A door gives access to the space under the seat, and when the tub is removed it is at once covered with a lid and placed in a van, while a clean tub, having in it a small supply of disinfecting fluid, is substituted for the full one taken away; each van holds twenty-four tubs.

The Goux system, which has been in use at Halifax, consists in lining the pail with a composition formed from the ashes and all the dry refuse which can be conveniently collected, together with some clay to give it adhesion. The lining is adjusted and kept in position by means of a core or mould, which is allowed to remain in the pails until just before they are about to be placed under the seat ; the core is then withdrawn, and the pail is left ready for use.

The liquid which passes into the pail soaks into this lining, which thus forms the deodorizing medium.

The proportion of absorbents, in a lining three inches thick, to the central space in a tub of the above dimensions, would be about two to one ; but unless the absorbents are dry, this proportion would be insufficient to produce a dry mass in the tubs when used for a week, and experience has shown that after being in use for several days the absorbing power of the lining is already exceeded, and the contents have remained liquid.

There would appear to be little gain by the use of the Goux lining as regards freedom from nuisance ; though it removes the risk of splashing and does away with much of the unsightliness of the contents, the absorbent increases the weight which has to be carried to and from the houses, and thus adds to the expense.

The great superiority of all the pail or pan systems over fixed middens, such as formerly prevailed, is due in the first place to the fact that the interval of collection is reduced to a minimum, the changing or emptying of the receptacles being sometimes effected daily, the period never exceeding a week ; and secondly to the receptacles for the refuse being independent of the structure, and therefore capable of easy periodical renewal when the material becomes saturated.

In the ordinary pail system, the plan of deodorising the material whilst the pail is in use is scarcely resorted to, from the trouble and expense it occasions ; and probably the best known contrivance for deodorising the excreta immediately after they fall into the receptacle, is Moule's Earth-closet, in which advantage is taken of the deodorising properties of earth. Many modifications of this system are in use.

Dry earth is a good deodoriser ; 1½ lbs. of dry earth, of good garden ground or clay, will deodorise each excretion.

A larger quantity is required of sand or gravel. If the earth after use is dried, it can be applied again, and it is stated that the deodorising powers of earth are not destroyed until it has been used ten or twelve times.

This system requires close attention, or the dry earth closet will get out of order, and the best dry closet is liable to a peculiar smell, which becomes very objectionable when the closet is not properly cleansed.

As compared with water-closets, the earth-closet is cheaper in first construction, and is not liable to injury by frost ; and it has this advantage over any form of cess-pit, that it necessitates the frequent removal of refuse.

On the other hand, the dry earth system is much more expensive than the pail system.

The dry earth system, though applicable to separate houses, or to institutions where much attention can be given to it, is inapplicable to large towns from the practical difficulties connected with procuring, carting. and storing the dry earth.

The pail system, which is in use in various ways in the northern towns of England, and in permanent camps to some extent at least, is more convenient as well as more economical than the dry earth system, where the removal of a large amount of refuse has to be accomplished.

In France the introduction of water-closets led to the adoption of *tinettes*, or metal vessels holding 50 litres each, which were placed under the seat, and furnished with a lid capable of being hermetically closed. These when full are removed by contract and the contents converted into manure.

The value of the manure depends on its bulk in relation to the distance it has to be carried. If the manure is concentrated, as is the case with *tinettes*, it can stand a comparatively high carriage. If the manuring elements are diffused through a large bulk of passive substances, the cost of the carriage of the extra, or non-manuring elements, absorbs all profit.

If a town, therefore, by adding deodorants to the contents of pails, produces a large quantity of manure, containing much besides the actual manuring elements—such as is generally the case with dry earth—when the districts immediately around have been fully supplied, a point is soon reached at which it is impossible to continue to find purchasers in consequence of the high cost of carriage.

According to theory, the manurial value of dejections per head per annum ought to be from 8*s.* to 10*s.* And if the pail system is to be commercially advantageous, the liquid and solid dejections must be collected without admixture of foreign substances.

General Scott, R.E., endeavoured to effect this by mixing the contents of the pails with lime, and distilling the mixture in a close boiler, so as to draw off four-fifths of the ammonia.

The ammonia thus evaporated was passed into a tank filled with sulphuric acid and sulphate of ammonia obtained in a crystallised state fit for the market.

The residue, after being mixed with superphosphate and dried, was sold as manure.

In a sanitary point of view the pail or tub system has an enormous advantage over the old midden or cess-pit system, in the facility with which the excrementitious matter is removed, without soaking into the ground or putrefying in the midst of a population.

There is no doubt that in some parts of India, where the water supply is often not plentiful, and the question of sewage removal presents many difficulties, a systematised pail system would afford great advantages.

Pneumatic System for the Removal of Excreta.

Instead of the removal of the excreta, as just described, by manual labour, there is in Paris, in Florence, in Philadelphia, and in other foreign towns, a pneumatic system in operation. The water-closet refuse is passed into close cess-pits, from which the contents are removed at frequent intervals. To effect this, in Paris a large cylinder, in Philadelphia a series of barrels, are brought on a cart ; a tube from them is connected with the cess-pit, the air is then exhausted in the barrel or cylinder by means of an air-pump, and the contents of the cess-pit drawn through the tube by the atmospheric pressure into the cylinder or barrels.

A plan which is practically an extension of this system has been introduced by Captain Liernur, in Holland.

He removes the fœcal matter from closets, and the sedimentary products of kitchen sinks by pneumatic agency. He places large air-tight tanks in a suitable part of the town, to which he leads pipes from all the houses. He creates a vacuum in the tanks, and thus sucks into one centre the fœcal matter from all the houses.

The mode of action is briefly as follows :—These tanks are in communication, by means of pipes, with small tanks in each street, so arranged that the vacuum in the central tanks can at will be extended to any given street reservoir.

Each of these street reservoirs is the centre of a small drainage system of houses, independent of all others, and the fœcal matter out of those houses is drawn into it by means of the vacuum, created in the manner described, after which the matter is at once dispatched to the main building by means of the same pipe that first conveyed the air. The closets are connected with a main pipe lying in the street by means of branch pipes, somewhat like the mains and branches of water-works. Each main has, however, only one stop-cock, and when this is turned all the closets connected with it empty themselves at once through the main pipe into the street reservoir.

The manipulation is as follows :—The air-pump engines maintain throughout the day a vacuum in the central tanks, and hence also in the tubes, extending the vacuum into the street reservoirs. A couple of men perambulating the town, visit each reservoir in turn once a day, and discharge into them, by opening the stop-cocks of the mains, the closet matter of all the houses belonging to the particular section connected with the reservoir ; after which, before leaving, they dispatch the whole to the central building by simply opening the stop-cock of the communicating pipe.

This system is extremely ingenious ; it is in operation at Dordrecht and Leyden, and has been experimented on at Amsterdam ; but from recent reports it does not appear probable that it will be largely adopted.

In a sanitary point of view it would seem prima facie to present some questionable features.

No current of air, however rapid, can be so rapid as the rate at which gasses diffuse themselves ; there must be always some film of filth left on the sides of the pipes, however perfect the vacuum ; it is moreover stated that in practice filth hangs about the pans of the closets.   Consequently, if any putrefying action went on in the pipes, decomposing matter or injurious gases might be liable to pass into the houses, if not in spite of the vacuum in the pipes and reservoirs, at all events when the vacuum is not in operation.

This method of removal of excreta necessarily requires a system of sewers, in addition, to carry off the water from baths, washing, kitchens, &c., and rain water.

The question resolves itself therefore into how most easily to restore this fouled water, to a condition fit to flow into streams and rivers without danger to the health of the community ; and, except in cases where special conditions prevail, it will be more convenient that the drains, which must be provided to remove waste water and water fouled in daily use, should also be employed to carry off the excreta and similar refuse which requires immediate removal from the houses.

CHAPTER XVII.

DRAINS IN A DWELLING.

THE removal from a dwelling of the water which has been polluted in the processes of daily life is a separate question from that as to how the refuse water is to be dealt with after removal.

The chief objects of a perfect system of house drainage are—

1. The immediate and complete removal from the house of all foul and effete matter directly it is produced.

2. The prevention of any back current of foul air into the house through the pipes or drains which are used for removing the foul matter.

The first object, viz. removal of foul matter, can commonly be best attained by the water-closet system when carried out in its integrity, though in certain cases a system of earth, or other form of dry closet, may be preferable. Cesspools in a house do not fulfil this condition of immediate removal. They serve for the retention of excremental and other matters, and they are inadmissible where complete removal can be effected. Where this is not possible, and cesspools must be had recourse to, they must be carefully designed ; they must be placed outside, and as far removed from the immediate neighbourhood of the dwelling as circumstances will allow.

R 2

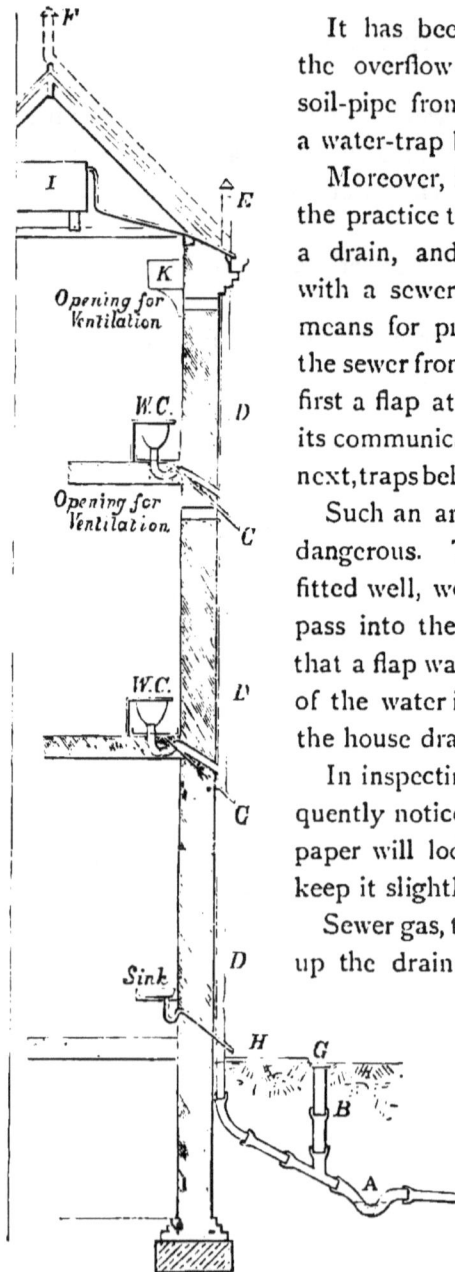

Fig. 50.

It has been the practice to carry the overflow from cisterns into the soil-pipe from the water-closet, with a water-trap between.

Moreover, it has also been largely the practice to carry the soil-pipe into a drain, and to connect that drain with a sewer; providing, as the only means for preventing the gases from the sewer from entering into the house, first a flap at the end of the drain at its communication with the sewer, and next, traps below the water-closet basin.

Such an arrangement is extremely dangerous. The flaps, even if they fitted well, would allow sewer-gas to pass into the house drain each time that a flap was opened by the pressure of the water in its effort to pass from the house drain into the sewer.

In inspecting sewers it will be frequently noticed that a bit of straw or paper will lodge under the flap, and keep it slightly opened.

Sewer gas, thus entering, would pass up the drain to the soil-pipe, where each time that a volume of water was thrown down, some of it would necessarily be forced through one or other of the traps, to make way for the inflowing water.

But independently of this, the capacity of water to absorb sewer-gas is very great, consequently the water in the trap would absorb this gas. When the water became warm from increase of temperature it would give out the gas into the house; when it cooled down at night it would again absorb more gas from the soil-pipe : and frequent change of temperature would cause it to give out and re-absorb the gas continually.

Hence if the trapped waste-pipe from a cistern is passed into the soil-pipe it will allow sewer-gas to pass through, and be absorbed by the water in the cistern : but many cisterns have a waste-pipe directly communicating with the soil-pipe, without any trap between.

Thus a water-trap without other precautions is but a frail protection against this very insidious and dangerous enemy, sewer-gas, unless it is diluted by means of a current of air in the pipe.

There is, moreover, another reason why fresh outside air should have free access to the soil-pipe. When a body of water is thrown down a closed soil-pipe it tends to draw in air, and is followed by the water from all the traps communicating with the soil-pipe, and thus leaves them empty.

To prevent these evils the following points should be attended to in house drainage :—

1. *General arrangement and position of drains.*

The first object to be attained in house drainage is to prevent the sewer-gas from passing from the main sewer into the house drains. The flap will not prevent it, and therefore a water-trap should be interposed. This water-trap should have as little surface as possible.

In large houses the break of connection with the sewer may be by a ventilated disconnecting manhole. But in small houses it may suffice to provide a syphon with pipe off the house drain, on the house side of the syphon, brought up to the surface of the ground, as shown in Fig. 50; in cases where this branch can be laid at an angle of about 45° it will not only act as a ventilator, but the syphon can be cleaned from it in case of stoppage.

As above mentioned the house drain, if closely sealed by water traps, will become a reservoir for the sewer-gas which may pass through this water-trap. It is therefore necessary in the next place to oxidize or aerate this gas. To do this we must ensure that a current of air shall be continually passing through the drains; both an inlet and an outlet for fresh air must be provided in the portions of the house drain which are cut off from the main sewer, for without an inlet and outlet there can be no efficient ventilation. This outlet and inlet can be obtained in the following manner. In the first place, an outlet may be formed by prolonging the soil-pipe at its full diameter and with an open top to above the ridge, in a position away from windows, skylights, or chimneys; and every branch pipe connected with a soil-pipe should be similarly carried up, and be left open at the top for ventilation. And secondly, an inlet may be obtained by an opening into the house drain, on the dwelling side of and close to the trap, either in connexion with the disconnecting manhole or branch pipe before mentioned, or by some other opening near the ground; or in special cases even by carrying up the inlet by means of a ventilating pipe to the level of the roof. The inlet should be equal in area to the soil-pipe, which should not as a rule be less than four inches in diameter. It may be assumed that if the inlet is low down, and if the outlet is carried to above the roof, there will not be any smell from the inlet; that is to say provided there is no deposit in the pipes.

But the value of these arrangements is lost, if the house drains be laid so as to allow of deposit; because if deposits occur in the drains or traps, they will putrefy and develop sewer-gas. But if the house drain and soil-pipe have a sufficient uniform fall to remove the refuse at once, and if they be well ventilated—and the more the openings the better—very little smell will arise. Indeed soil-pipes and drains cannot be too open or too much ventilated, provided always that they are cut off by a water-trap from the main sewer, which should always be ventilated also.

Flues in walls should not be used for sewer or drain ventilation; the sewer-gas would be liable to permeate the wall and pass into the house. All pipes or openings for ventilation should be external. Ventilating pipes should be carried upwards without angles or horizontal lengths, and with air-tight joints. The upper end should not be near windows or ventilating openings, and should be carried above the level of the ridge of the roof.

The best position for a soil-pipe is outside the house; because any escape of sewer-gas resulting from defective joints, corrosion, or otherwise would take place in the open air.

In cold climates, or exposed localities, pipes so placed would require protection from the frost.

The mean temperature of the internal air in town sewers in this country from a year's experiments was found to be about 12° higher than the external air in winter, and 3° cooler than the outer air in summer. The temperature of sewage is always above freezing-point, and in this country danger from frost might easily be guarded against.

Waste-pipes from baths, lavatories, or sinks (except slop sinks for urine or water-closet sinks, such as are used in hospitals), must not pass direct into the drain.

Similarly, waste-pipes from cisterns or rain-water tanks must not communicate with drains. In a well-arranged set of cisterns in a house, the overflow from the upper cistern will fill the lower ones conclusively, so that one waste from the lower cistern would suffice, and this one should deliver into the open air. Moreover, to prevent the possibility of the drinking water becoming polluted by sewer-gas, which might find its way up the pipe which supplies water to the closet-pan, each water-closet should have its own special cistern supplied by a ball-cock from the principal cistern, and it should not be possible to draw water from any cistern supplying a water-closet for any other purpose than the supply of such water-closet.

It should therefore be an absolute rule that no overflow or waste-pipe from any cistern or rain-water tank, or from any sink other than a slop-sink for urine, or slop-sinks for hospital use, or from any bath or lavatory, or safe of a bath, or of a water-closet, or of a lavatory, should pass directly to any drain, soil-pipe, or trap of a water-closet ; but every such pipe should pass through the wall to the outside of the house and discharge near or over trapped gullies with an end open to the air, or deliver into a pipe which so passes and discharges. But all waste-pipes should be properly trapped close to the sink or lavatory or bath, because any pipe through which fouled water passes at intervals will have its interior coated with deposit, which will occasion a smell. This smell will be removed by fresh air, therefore a current of air should be maintained in all waste-pipes which are open at the bottom, by means of an opening at the upper part as far away from windows as can be conveniently managed. Care must be taken to protect waste-pipes against frost.

Sinks and water-closets should be placed against external walls, so that the pipes conveying refuse water or soil may be more immediately passed outside the main wall, and the rooms or closets so appropriated should have windows opening directly to the air.

Water-closets or sinks situated within houses which are so placed as to have no means of direct daylight and external air-ventilation, are liable to become nuisances, and may be injurious to health ; and if such sinks and water-closets cannot be ventilated in an efficient manner they had better be removed.

2. *Constructional requirements in laying house drains.*

Every drain should be laid in a straight line with proper falls and true gradients. All drains and soil-pipes should be round and not angular. Round pipes are more self-cleansing than angular pipes, and the circumference being smaller in proportion to the area, they cause less friction in the passage of the fluid.

Drain-pipes as distinguished from soil-pipes and other down pipes, are generally (1) of glazed stone-ware, with asphalte joints for inside work and cement joints for outside work; (2) of concrete cemented inside and with cement joints; or (3) of cast iron pipes coated inside with tar, with lead water-tight joints. Lead joints can only be made in a strong iron pipe; and the use of these joints is to some extent a guarantee of soundness, but every iron pipe should be carefully tested by water pressure to see that there are no holes or flaws.

The use of cast iron for house drains, if the cast iron is solid, sound, and free from porosity, will prevent leakage and subsoil tainting beneath the house, and will be as cheap as earthenware pipes in many cases.

In laying pipes in the ground, sufficient space should be cut out in the ground at the joints to allow of a good joint being made, and examined before being closed in ; to ensure accuracy in laying, the pipes for house drains should always be bedded on concrete. No right-angled junction should be allowed, except in the case of a drain discharging into a vertical shaft. It is advisable to construct access channels with manholes at every change of direction, to admit of examination of the drain.

No drain should pass through and under a dwelling-house, except where absolutely necessary. When it must so pass, it should be laid in a straight line under the house, between disconnecting manholes placed outside the house, arranged for cleansing and examination ; the portion of the drain which is under the house should be of cast-iron pipes laid in an air drain, so constructed as to be as impermeable as possible, open to the air at each end, and allowing access to the drain-pipes when necessary. Where possible, all connections with drains should be made outside a house.

In laying the drain-pipes proper junctions must be provided in all drains for sink and water-closet drainage, and for drain ventilation ; because communications with drains without proper junctions lead to deposit and leakage.

For an ordinary house a 4-inch drain-pipe is ample—a 6-inch pipe would be required for a large hotel—a 9-inch pipe would suffice for a very large building. To ensure freedom from deposit, pipes for moderate-sized houses should never be larger than six inches.

A fall of 1 in 40 is a good fall for a 4-inch drain,

,.	1 in 60	:,	,,	6 ,,
,,	1 in 80	.,	,,	9 ,,

If from insufficient fall or other causes the drain cannot be kept clear of deposit without flushing, special flushing arrangements must be provided. Leaving taps open, and propping the handles of water-closets, will not flush drains, but will only waste water. A large flush of water is required for flushing drains. Where water is plentiful, a discharge direct from a large cistern provided for the purpose may be obtained. Where water is limited, the drains may be flushed by introducing small paddles in grooves formed at the access or junction chambers, and suddenly releasing the pent-up water; or where the flow of sewage is small, a tumbler flushing box or Field's self-acting flush-tank may be used.

Mr. Roger Field's flush-tank is designed for flushing house drains, but may of course also be used for flushing sewers. The syphon is so constructed as to be put in action by a very small constant flow of water. This is effected by making the discharging limb of the syphon and its connection with the bend of the syphon in such a way that when a small quantity of water flows over the bend, it falls clear of the sides, instead of running down along the sides of the discharging limb. Figs. 51, 52.

By this arrangement, the water as it drops carries away the air in the discharging limb, and thereby starts the syphon. The longer limb of the syphon must be dipped into water below the tank. A convenient arrangement for this purpose is an annular form of syphon, as shown in the drawing. By this means a large syphon can be put in action by a very small flow of water, so that a large flushing tank may be fed, for instance, by the constant flow from a small watertap which would take a day or two to fill it, and yet the tank will discharge itself automatically as soon as it is full.

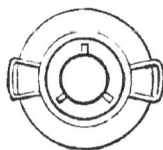

Fig. 51.

The syphon is capable of being easily taken to pieces in case of any stoppage occurring. The outer case can be readily lifted off, and as readily replaced for the purpose of examining or cleansing the syphon.

Rain-water pipes should deliver into an open channel or over a gully, or in some other way so as to have the discharging end open to the air. In special cases they may be connected with the drain so as to be used as ventilating pipes, provided their upper extremities

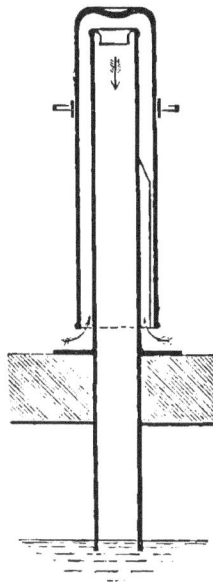

Fig. 52.

are situated at such a distance from windows, openings, or projecting eaves, as to ensure that there is no danger of escape of foul air into the interior of the house from the pipes. When so used the joints should be air-tight.

All inlets to drains should be properly trapped, except where left open for ventilation of the drains.

The gullies to receive discharge from waste-pipes or rain-water pipes, surface drainage or otherwise, should be trapped (see Figs. 53, 54) : they should be placed at a distance of a yard at least from windows. The waste-pipe from a sink may discharge under a grating for appearance sake.

The surface of water exposed in a trapped gully should be as small as possible, compatible with facilities of cleansing, so as to limit the surface for evaporation. The gully at which the waste-pipe from a scullery sink discharges should be provided with a grease-trap outside the house ; otherwise grease may choke the drains.

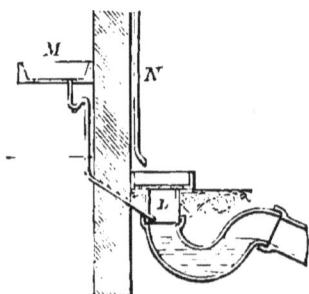
Fig. 53.

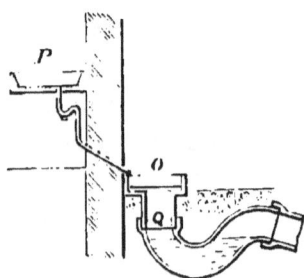
Fig. 54.

Gullies should not be placed inside a house, in cellars, basements. or otherwise, unless absolutely necessary. Where such gully cannot be avoided, it should be properly trapped, and the outlet pipes should not pass directly to any drain or sewer, but should be disconnected therefrom by passing through the wall to the outside of the house, and there delivering with an end open to the air over a suitable trap or in a disconnecting manhole (Fig. 55). Where the basement is so deep that this could not be arranged. it would be advisable to run the drain to a small brick and cemented well-hole to be pumped out daily, or when occasion might require, so as to avoid direct communication with a drain.

When, from the dampness of a site, it is necessary to lay subsoil or land drains, and to discharge the water into a sewage drain, the subsoil or land drain should not pass directly to the sewage drain or sewer, but should deliver into a disconnecting manhole, as in Fig. 55.

All inlets to the drains and all openings for ventilation should be efficiently protected by gratings, to prevent the introduction of improper substances. When not so protected the gully should be arranged so that stones, sand, or other heavy matters falling in will drop to the bottom clear of the trap and entrance to the drain ; and the gratings should be moveable, to allow of such heavy matters being easily removed from the bottom of the gully, and they should be cleaned out periodic-ally.

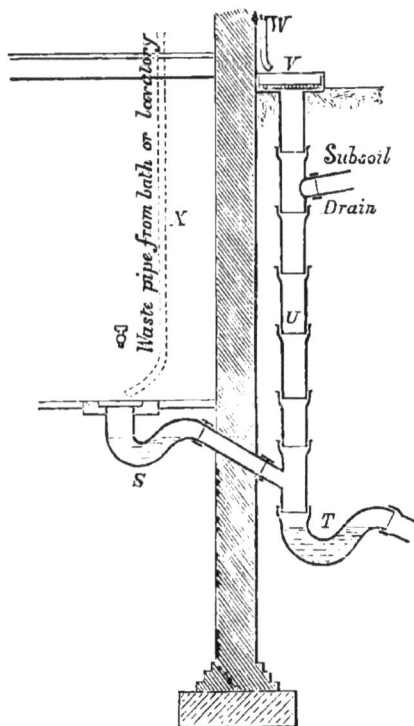

Fig. 55.

Gratings to inlets, or openings used for ventilation or disconnection, should have a free air space at least equal in area to that of the drain-pipe or gully at the place.

Gratings should be made of a form to allow the water to pass freely, and to prevent dirt from lodging and choking the passage for the water. Circular holes in gratings are least adapted of any form for allowing water to pass.

3. *Soil and waste pipes.*

Soil pipes carry away the discharge from water-closets and from slop-sinks.

A 4-inch diameter soil-pipe is sufficient for ordinary-sized houses; if many closets are on the same pipe then a 4½-inch or 5-inch may be necessary, very rarely a larger size. Solid drawn lead pipes are preferable. 6 lbs. lead is the least thickness admissible. 7 or 8 lbs. lead, and more, with the larger pipes, should be used in all cases where hot water is occasionally thrown down, being better calculated to resist the alternate expansion and contraction.

Iron pipes are sometimes used, but the making of a perfect joint is more certain with lead than with iron; and cast iron, if used, must be sound, and free from porosity. Joints in iron pipes should be made with lead. Cement joints cannot be relied on, either for lead or iron pipes, as they may crack. The down pipe, whether lead or iron, must be properly supported at intervals, so that its weight or the jar of the flow of water shall not tend to move it and open the joints, either at the junction with the closets or at the bottom, and thus allow of a leak of sewer-gas.

Lead soil-pipes are usually supported by blocks at every ten feet, but intermediate supports by means of a flange or a tack soldered to the pipes are desirable. Soil or down pipes placed inside a house should be placed in a space built in the wall, of sufficient size to allow of the joints being well made and examined. If placed in chases cut in the wall, the sides of the chase should be carefully cemented and smooth. Casings made to open should be provided to the chases. But it is far preferable that where possible the soil-pipes should be entirely outside the house; so that if a leak should occur, it will not do much harm; because there is always danger of concealed lead pipes being injured by nails being driven into them inadvertently. Outside soil-pipes require protection from frost.

The connection with the drain should be outside, because

the junction of the vertical soil-pipe with the house drain is a place at which there is always risk of cracks, which would allow of the leakage of sewer-gas. On this account it is advisable when a lead or an iron soil-pipe must terminate in an earthenware drain-pipe inside a house that the bottom of the soil-pipe should discharge into a trap in the earthenware pipe. In such a case the horizontal drain and the soil-pipe should each have independent adequate ventilation by an inlet and outlet for air.

The waste from a cistern should be of a size capable of emptying the cistern very rapidly from its lowest point, so that the cistern may be easily cleaned. Cisterns should be thoroughly cleaned out at least once a month. The water from the cistern should not discharge directly over or into a gully, for fear the trap should be dry and foul air pass up the waste into the cistern.

Waste-pipes from baths, as well as the valves, should be at least 2 inches diameter, and from basins at least 1½ inches diameter for comfort. Lead waste-pipes carrying away hot water, should be of at least 8 lbs. lead. Where branch wastes communicate with a main waste-pipe, bends in the branch waste are desirable to allow for the expansion and contraction of the wastes.

4. *Fittings for water-closets.*

Many kinds of water-closet apparatus and of so-called 'traps' have a tendency to retain foul matter in the house, and therefore in reality partake more or less of the nature of small cesspools; and nuisances are frequently attributed to the ingress of 'sewer-gas' which have nothing whatever to do with the sewers, but arise from foul air generated in the house drains and internal fittings.

The ordinary form is a basin with a hole below, which is sealed by a moveable lower basin, which holds the water. A lever movement is required to drop the second basin. so that its contents may pass into the drain: but, in order to prevent the sewer-gas from passing through the hole in the first basin at the moment of discharge of the contents, and up into the closet, and also in order to prevent the sewer-gas from passing, as it would do continuously, round the lever apparatus into the closet, the contents of the moveable or second basin are dropped into a receptacle with a D trap.

This trap is out of sight, and it is necessarily rather large, and from its shape it is liable to accumulate much foul putrescent matter. To remedy this hundreds of forms of water-closet pans have been introduced, of more or less merit.

In the selection of a pan, the object to be attained is to take that form in which all the parts of the trap can be easily examined and cleaned, in which both the pan and the trap will be washed clean by the water at each discharge, and in which the lever movement of the handle will not allow of the passage of sewer-gas.

The trap is, however, useless if any escape of gas can take place at its junction with the waste-pipe; consequently the trap should always form a portion of the soil- or waste-pipe, instead of a part of the fitting, and the water-closet or sink should be capable of being moved without disturbing the trap. Lead is the best material in a house for traps and waste-pipes, because it is smooth, durable, free from corrosion, and joints can be made easily and safely in lead. For outside work stoneware may be used. The best form of lead trap is a smooth cast lead trap without corners. When such a volume of water is thrown into a trap as completely to fill it, the trap will act like a syphon and empty the pipe, leaving the trap without water; consequently it is desirable to have a small ventilating pipe at the bend beyond the trap carried into the main ventilating pipe or into the open air, to prevent the syphoning action. Lead traps through which hot water passes should be of 8 lbs. lead. No trap should be of less than 6 lbs. lead.

The depth of dip of a trap varies from half an inch to as

much as 3½ inches. It should depend on the frequency of use. When rarely used more depth of water is required. to prevent the trap failing from evaporation. Openings for cleaning traps should be below the water-level of the trap, then they show by a leak if they are not tight.

5. *Cesspits.*

The arrangements described have special reference to the case of sewers; but in certain cases, such as private houses, or small detached barracks, it is convenient, and indeed often necessary to provide a cesspit, to receive and retain the refuse for a certain time.

The old form of cesspit was to dig a hole in the ground ; if porous ground, so much the better; and to pour into this the whole foul refuse of the house or barracks, until its over-flow compelled its cleaning out. Numerous cesspits of this sort used to exist in barracks, which the porous soil had so favoured that there had been no necessity to empty them out for many years. These cesspits consequently saturated the soil and polluted the adjacent wells.

From what has been said of the movement of ground air, it is apparent how dangerous this mode of dealing with refuse must be, especially if the cesspit is near the house. Of course the risk of danger to a house from this pollution of ground air from the cesspit will increase with the depth of the cesspit. But in addition to the risk of pollution of the ground air, there is the risk of pollution of water in adjacent wells.

If it is necessary at any place to turn the refuse into cess-pits, the cesspits should be made impervious by puddling outside, and lining with brickwork in cement ; they should be as small in size as circumstances allow, so as to necessitate frequent cleansing.

It is essential that the gases generated in the cesspit should not pass into the house drain.

The cesspit must be covered, but if necessarily near a

S

dwelling. it should have ventilation through a large charcoal filtering screen, carefully protected from wet; if away from dwellings, large openings protected from rain would suffice; and there should be a water-trap between the cesspit and the house; and between this trap and the house, the ventilation of the house drains should be arranged on the principles already described, so as to ensure that any gases which pass the trap may have an opportunity of being diluted with fresh air.

When cesspits, or sewage filters have to be provided for barracks, as frequently occurs in the new Depot Brigade Barracks, they should invariably be outside the barrack enclosure, and removed from the barracks as far as the ground in possession of the Government will allow.

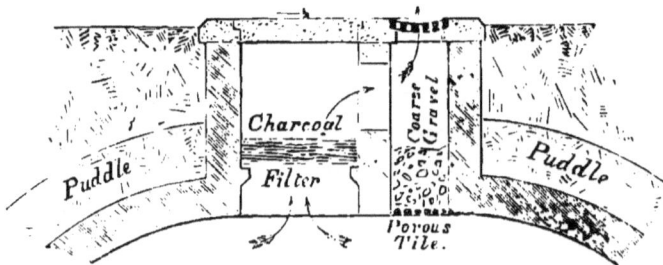

Fig. 56.   Ventilator for Cesspool.

In these suggestions it is assumed that the sewers or cesspits from which the house drain is cut off, are themselves adequately ventilated, for otherwise the precautions suggested would be much less effective.

6. *Examination of drains.*

It will be seen from these few remarks that in the question of house drainage safety depends quite as much on the workman who executes the work as on the architect or engineer who designs it; and this part of the subject would not be complete without a few observations on the examination of house drains.

In examining a drain the main drain should be opened at

the point where it leaves the house, and by pouring water down the various closets, sinks, bell-traps, and rain-water pipes, the various pipes may be traced, and some opinion formed as to the state of the drains. The velocity of the water flowing in the drain may be ascertained by pouring water down at any point, and accurately noting the time it takes to reach the opening. After examining the basement, every closet, sink, bath, &c., should be examined in detail. To do this it is necessary to have the wooden casings, seat of closets, &c., removed, to trace how each closet, sink, bath, &c., is supplied with water, how the soil-pipes from the closets, and the waste-pipes from sinks, baths, overflows of cisterns, waste to safes, &c., discharge. It may be roughly determined whether the drains are trapped off from the main sewer by ascertaining if there is any draught from the bell-traps in the sinks. This mode of examination may be helped by pouring ether down a soil-pipe, when the fumes will be perceptible in the house at any leaks in the soil-pipe or failures in the traps. But in order to make a complete examination of the condition of drains and soil-pipes in a house, it is necessary to resort to the sulphur and the smoke test. Sulphur fumes may be applied by putting into an opening made in the lowest part of the drain an iron pan containing a few live coals, and throwing one or more handfuls of sulphur upon the coals, and closing up the opening to the drain with clay or otherwise. The fumes will soon be very perceptible at any leaks or rat-holes in the soil-pipe, drains, or traps.

The smoke rockets sold by all makers of sanitary appliances will afford a still more efficient test.

A length of sewer pipe may be conveniently tested by filling the length with water for a specified distance and ascertaining whether the water leaks away. But this method requires caution, and it is unadvisable to use a head of water of above 2 or at most 3 feet, as the pressure of a greater head might prove too severe for the joints.

CHAPTER XVIII.

WHATEVER may be the system adopted for the removal of
the excreta, there will always be a large amount of liquid refuse
to be dealt with which is liable to putrefy, and consequently,
in providing for the removal of this, it is generally more
economical to arrange at the same time for removing the
excreta, than to institute a separate service for each.

In the removal of the foul water, the question is often
mooted of providing for the water used in households only,
and making a separate service for rainfall.

This question is one which has obtained considerable
importance from the theoretical views which have sometimes
been permitted to decide it.

A strict rule cannot be laid down.   In the case of a detached
house or of a barrack, or of a small collection of houses, the
conditions are very different from those in large or even
moderate sized towns.

In the case of detached houses and barracks, it is frequently
found advisable to save the water which falls on roofs, and to
use it for washing and other purposes requiring soft water.
But when, as in a large town, the ground is closely occupied
by houses, the rain-water, as it falls, is fouled by soot; and it

moreover takes up many other impurities from the atmosphere, and is unfit for domestic use.

In a town area, whilst the soot and the impurities of the atmosphere which the rain-water takes up in falling frequently render it unfit for domestic purposes, that which falls on streets and yards is made quite foul by the large quantities of horse-dung and other impurities which are the chief constituents of the street dust in towns.

Experiments show that, whilst the solid matter in rain-water which passes into sewers from wood pavement consists to some extent of wood fibre, and that a small part of the solid matter from paved streets consists of granite, the bulk is horse-dung ; hence the value as manure of street water taken in rainy weather on its way to the sewers is as great, or even greater, than the ordinary contents of the dry weather sewage.

It is on these grounds necessary to provide in the drainage of a large town, as far as possible, for the removal of the foul street and yard water, as well as the house sewage proper.

If provision were made for the very large amount of rainfall which may be expected to occur exceptionally and at rare intervals, other elements of inconvenience and even danger to health would be introduced. For instance, the small amount of the dry weather flow of sewage would necessarily only form a trickling and stagnating stream in a sewer calculated to remove the maximum volume of rainfall, and this dry weather flow would be liable to create deposits. A large sewer, in which deposits can occur, would form a great reservoir for sewer-gas.

Where provision is made for rainfall, the nature of the area which has to be drained must regulate the proportion of rainfall to be removed. For instance, in a town area, closely built over, the rain-water will run off rapidly into the drain ; in a suburban area, that which falls on garden ground will pass off more slowly. The rain-water which passes from streets and courts into a sewer during a storm immediately succeeding a dry season, is generally richer in manurial value than that which passes in after the rain has lasted some time.

But in any town or district sewered effectively, in which the admission of surface water is limited, and in which the subsoil waters are excluded from the sewers, the volume of sewage will be reduced, and be thus more easily dealt with, either by pumping or by gravitating to land for irrigation.

The excess of rainfall not provided for in sewers must be carried off by carefully formed surface drains; or else by storm overflow channels formed in the sewers to pass the water into natural watercourses: to effect this object many ingenious devices have been designed.

The flow of water from a drainage area during continued heavy rain, as in a wet season, or during a thunderstorm, may swell the rivers and natural streams so as to swamp and water-log the lower parts of the sewers. When the outfall is in a tidal estuary or in the sea, tides are liable to backwater the outlet-sewers. Some tidal rivers in England rise vertically in flood as much as 20 feet. In some of these cases pumping is resorted to ; in other cases the sewage is retained in tanks. or in tank-sewers, until the ebbing of the tide, and no absolute rule can be given when one or the other mode shall or shall not be adopted. Where an outlet sewer is liable to be back-watered, it may sometimes be advisable to keep the invert up as much as possible, even at the expense of the gradient, and to depend on flushing.

The question as to whether this plan should be adopted, or whether on the other hand the invert should be kept down, although liable to be backwatered daily, will depend upon whether a sufficient velocity of flow can be obtained during the period of maximum discharge to remove any silt which may have been deposited during the stagnation arising from the backwater.

The velocity of flow of water which will move materials is regulated by the following considerations :—(1) For objects of the same character, the velocity required to start them increases with the mass of the object; (2) for different objects, the velocity required increases with the specific gravity; (3) according as the object assumes a form approaching a sphere, the less velocity it takes to move it; whereas a flat object, like slate, requires a considerable current before it becomes disturbed from its position; (4) the object moves at a less rate than the current, but when the velocity of current increases after an object is in motion, the velocity of such object increases in a progressive ratio.

A velocity which will start an object will (when constantly maintained, and no accidental circumstance occurs to prevent it) never allow such object to deposit in the stream.

A velocity of from 2 feet to 2 feet 6 inches per second with a continuous flow will remove all objects of the nature of those likely to be found in sewers; but if the flow is intermittent, or if a part of the sewer remains occasionally stagnant, a velocity of 3 feet would be advisable during a portion at least of the period of maximum flow.

Sewage should not be allowed to acquire a velocity of more than four feet per second. Six feet per second will move grit and other solids along the sewer invert, with a cutting action rapidly destructive of the material of the sewer.

The sewage of a town or village will consists of waste-water and excreta from the houses, and the volume, in round figures, may range from 100 to 250 gallons per day from each house.

This flow of sewage is not uniform throughout the day. It varies with different places, according to the habits of the people. In London, 46 per cent. of the daily dry weather flow reaches the outfalls between 11 a.m. and 8 p.m., whilst only 21 per cent. flows off between 2 a.m. and 9 a.m. The largest flow per hour amounts to 6 per cent., and the smallest to 2·6 per cent. of the whole daily dry weather discharge. The volume of maximum flow regulates the size of the sewer, so far as sewage is concerned.

In addition to this, the provision for rainfall has been assumed in London at ¼ of an inch of rainfall to be carried off in 24 hours from the urban districts, and ⅛ of an inch in 24 hours for the suburban districts.

As all rainfall cannot be excluded from the sewers, it has been suggested to provide for a total amount of not less than 1000 gallons from each house; that is to say, for a town of 1000 houses (5500 population) the sewers should have a delivering capacity of about 1,000,000 gallons. An outlet-sewer of 2 feet diameter, laid with a fall of 5 feet per mile, will deliver upwards of 2,000,000 gallons, flowing a little more than half full; and as provision should be made for increase of population, a sewer of 2 feet diameter may be provided for each 5500 persons, where no better fall than 1 in 1000 can be obtained. Lesser diameters will answer where there are greater falls.

A comparatively long line of outlet-sewer may be necessary to intercept and take the sewage of several lines of sewers to some common outlet, where there are sewage tanks, or 'sewage-farms'; this outlet may, however, be confined to, say, three times the ordinary flow of sewage, if storm-water overflows can be provided; because it would not as a rule be desirable to construct the intercepting sewer of sufficient capacity to remove storm-water, as neither sewage-tanks nor sewage-farms can deal satisfactorily with such excessive volumes of water. In some cases it may be necessary to provide a storm-water sewage-tank; for this purpose an area of low-lying land may be deep drained and embanked, and the excess of sewage during rain be discharged on to it and left to filter away in the intervals. Such an area would be a rough sewage-filter, to be used during wet seasons. A portion of land on the margin of a watercourse may, in some cases, be embanked above flood-level for this purpose.

In laying out the sewerage of a town, it is of importance that the water from the highest levels should be carried off independently of the lower levels, when it can be done.

Gibraltar afforded a striking instance of this (Fig. 57). The town is at the foot of the rock—part is situated on the steep side of the rock, and part on a flat piece of land, which is separated from the sea by the fortifications.

The sewers were carried down the steep streets, which ran from the upper part of the town, across the flat portion of the town, and through the line wall into the sea. One of the main sewers commenced at a height of about 128 feet above high water mark, and reached the sea-level after a course of about 1000 feet.

Another commenced at a height of 264 feet, and descended to the sea-level after a course of about 1500 feet, 500 feet of the lower end being nearly horizontal. A third had a fall of

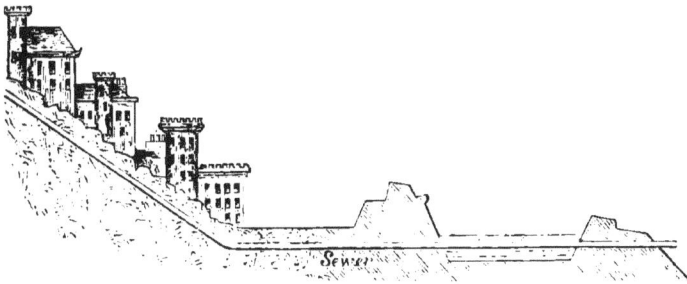

Fig. 57. Former Sewers at Gibraltar.

264 feet in 2200 feet. A fourth discharged its water into a nearly horizontal main, after a fall of 232 feet in 1550 feet.

Therefore the rainfall on the surface of the town, and on the roofs of the houses, together with the deluge of surface water which descended from the highly inclined slopes of the rock above, ran with extreme rapidity from all the higher levels into the lower and flatter districts, where, its velocity being suddenly checked, it used to burst the sewers, saturate the subsoil, and interfere with the efficient drainage of all the lower parts of the town.

The rapid current down the steep part became a slow

stream in the flat portion, deposited a large sediment, and in dry weather became a source of intolerable nuisance. This has now been remedied by intercepting sewers, which carry the water from the upper part of the town to the sea, independently of the lower levels.

Similarly the low-level districts in London are protected from flooding by means of intercepting sewers.

Those portions of any system of sewers which are below the level of high water of the sea or of the land floods of an inland river, should be cut off from the upper portions, and each portion should be provided with a separate outlet.

The portions of a sewerage system which are below the level of high water, or of floods, must in some cases have the sewage lifted so as to obtain a means of discharge when the outlet is submerged.

Where there are two or more of such low-lying places separate from each other, it may be economical to derive the power from a central pumping station, and to transmit the power by automatic arrangements to the other places where it is required. The methods of automatic transmission of power which have been particularly adopted for raising water or sewage are water-pressure, air vacuum, or pressure, and electricity.

The last-named is still in a somewhat experimental stage. Air in a vacuous condition has long been used for raising water by drawing it into air-tight receptacles, which when full deliver it automatically at the higher level, a succession of vessels being employed in cases where the lift is in excess of that which could be overcome by the pressure of the atmosphere.

Air under pressure has been employed to drive engines at a distance to work sewage pumps. This is a mode of using compressed air which admits of its being employed expansively, and is therefore economic as regards power, and as the air pressure may be high and not dependent upon the 'head' of sewage to be lifted, the mains may be comparatively small.

There is another method of using compressed air for raising sewage, where, as in the vacuous method above referred to, for raising water, pumps are not employed, but where an air-tight receptacle is provided, into which the sewage of the district falls, and is collected until the receptacle is full, when, automatically, the inlet valve closes, and immediately after an air-valve is opened which allows air under pressure to enter, and to exert that pressure on the surface of the sewage, driving it out of the reservoir, through the delivery pipe to the sewer at a higher level, into which it has to be delivered.

Lastly, there is the mode of transmitting power by water under considerable pressure, the pressure commonly employed being from 700 lbs. to 800 lbs. to the square inch. This is a system which has to recommend it the experience of its thorough and efficient working on a large scale for more than a third of a century. Not only has it this long and satisfactory record, but its employment increases from year to year, so that, at the present time, it is to be found doing the whole of the cranage work in almost all large docks, driving capstans and other engines in these docks, and at railway goods stations. So very beneficial has its employment been found that it has become worth while for a company to establish itself in London for the purpose of supplying this high pressure water power to private consumers. It has recently been adopted in preference to other systems for emptying the mines of water in the South Staffordshire district.

There is no mystery of complexity about working such a sewage pump, and the mode of transmission of power hydraulically is found on the score of its efficiency, of its economy, and its freedom from any collateral disadvantages to be the best for use in automatic sewage pumping stations.

The intercepting system is especially needful in towns

where the levels vary much, because a sewer which is carried down a steep hill becomes a chimney up which the sewer-gas flows, and may pass through the drains into the houses at the upper part. Consequently, when a sewer is necessarily placed in such a position, the upper or dead end beyond the junction of the highest house drains should be kept open, and numerous openings should be provided in its course to allow of its complete aeration.

Means for full and permanent ventilation of town sewers and house drains are required to prevent stagnation or concentration of sewage gases within sewers and drains. Ventilation requires both inlets and outlets : these will be adequately afforded by making numerous openings from the sewers to the external air, the object being to cause unceasing motion and interchange betwixt the outer air and the inner sewer air, which will bring about and maintain extreme dilution and dispersion of any sewage gas so soon as generated.

In unventilated sewers the concentrated gas becomes deadly, whilst in fully ventilated sewers with continued flow without deposit, the sewer air is purer than that of stables, or even than that in a public room when fully occupied.

Ordinary main sewer ventilation should be provided for all sewers, at intervals not greater than 100 yards ; that is to say. not fewer than 18 fixed openings for ventilation should exist on each mile of main sewer.

If sewer air at any sewer ventilator, or at any other point, should be offensive, additional means for prevention of deposit and for ventilation on this sewer are required, and should, as soon as possible, be supplied.

In properly constructed sewers the ventilation should prevent any smell ; in old sewers of defective construction charcoal filters for the ventilators have sometimes been adopted. Charcoal retards the ventilation, and it should only be looked on as a temporary makeshift until the sewer has been reconstructed or adequate flushing provided so as to avoid deposit and to ensure free interchange of air between the sewer or drain and the outer air.

For detached houses, villa residences, or larger establishments, drains should never end at the house which has to be

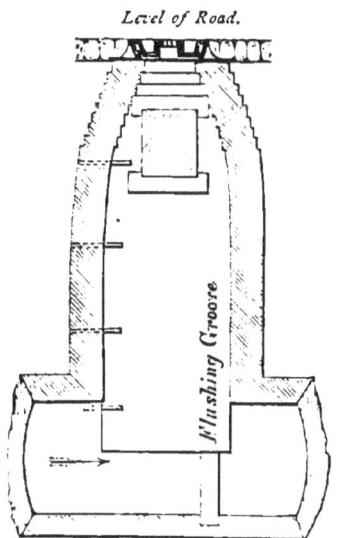

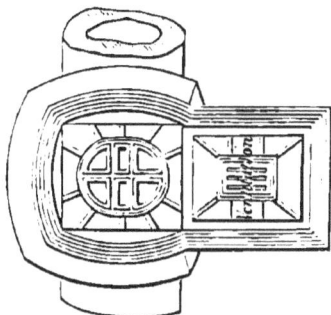

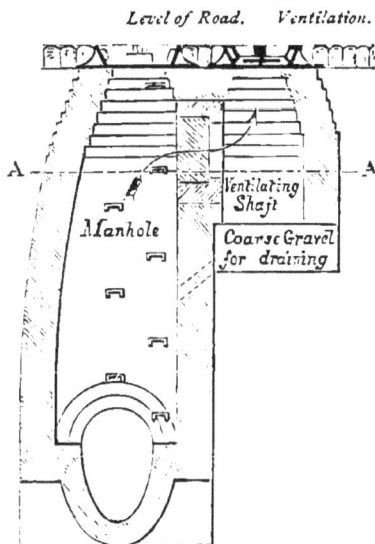

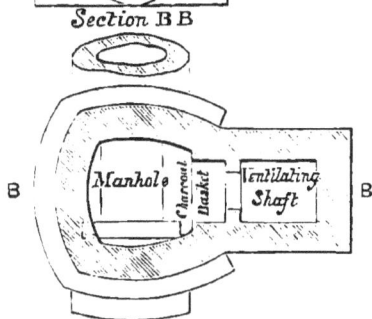

Fig. 58.   Plan at top.   Fig. 59.   Plan at AA.

drained, but should be continued beyond and above to some higher point or ventilating shaft where means for full and permanent ventilation can be provided, so as effectively to relieve the house from any chance of sewage gas contamination.

With the abundant means for ventilation suggested, the air within the sewers will be made comparatively pure by dilution and oxidation, and the further dilution and dispersion which would occur when the gas passes into the open air will generally dissipate remaining traces of taint and danger.

Manholes should have moveable covers at the surface of the ground. There should be a side chamber for ventilation, ' step irons,' to give access to the invert, and a groove in the invert and sides to allow of a flushing board being inserted at will for flushing purposes. Details for manhole and side chamber for sewer ventilation are given in Figs. 58, 59.

The ends of all sewers and drains at the lowest outlets must be so protected that the wind cannot blow in and force any sewage gases back to the streets and houses. Flap-valves, or other contrivances, may be provided to cover and protect outlet ends of sewers and drains, and so prevent the wind blowing in ; or the mouth can be turned down into the water.

Where the flow in the sewer is uniform in quantity, or where the minimum flow is equal to half the maximum flow, a circular form of sewer is best, if made of such a size that the depth of water in the sewer at the time of maximum flow shall afford the maximum hydraulic mean depth (i. e. the quotient afforded by the area of section divided by the wetted perimeter) ; but when the flow in a sewer varies very considerably at different times of the day, the egg-shaped form of sewer is preferred as affording better results with a small flow of sewage. Figs. 60 and 61 show the form now generally adopted. Fig. 62 shows the section of the river at Brussels, where it is covered over, and the intercepting sewer which runs parallel to the river. The sewer in this case is formed with a smaller trough at the lower part and expanded at the upper part, by which means a higher velocity is obtained with the dry weather flow of sewage than would otherwise prevail, and the larger area provides for carrying off rainfall ; during the dry weather flow a footpath is reserved on each side. When the sewer is quite full the overflow passes to the river through side flaps, which the pressure forces open.

Steep gradients in sewers must be modified by forming vertical steps, or falls, to prevent the sewage, during heavy rains, from acquiring such a velocity as shall not only wear out the invert and blow the joints, but also burst the sewers.

Old Form.   Improved Form.

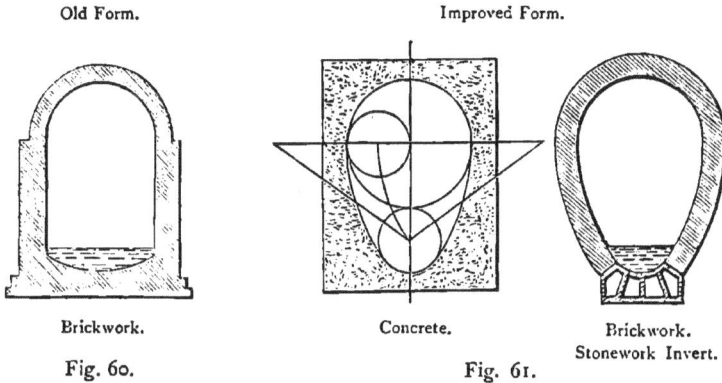

Brickwork.   Concrete.   Brickwork.
                                       Stonework Invert.
Fig. 60.   Fig. 61.

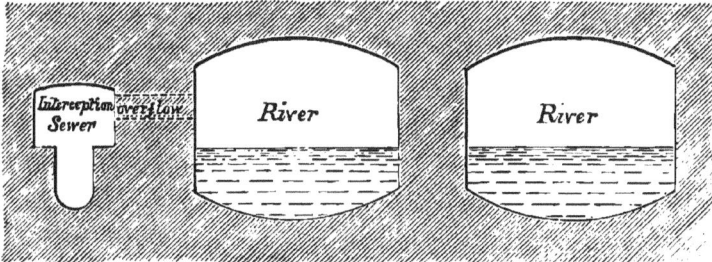

Fig. 62. River and Sewer Conduits at Brussels.

The steps, or falls, should be so formed as to prevent any accumulation of deposit. Earthenware pipe sewers, when laid down steep gradients, should be bedded and jointed with concrete.

Sewers should be watertight. If the water finds its way out of a sewer, the diminished flow causes sediment to be deposited, and obstructions to be created, and the subsoil round the sewer becomes polluted with sewage matter, consequently in dry porous subsoils the trench cut for the sewer should be made watertight with clay puddle.

Brick sewers should never be set dry to be grouted—they should be set in hydraulic mortar. If sewers are to be watertight so as to exclude subsoil water, the bricks must be set in Portland lime mortar, with a lining or coating of mortar outside the sewer, and betwixt each 4½-in. course of bricks. Wet subsoils may require special land drains. Cast-iron pipes may be used for main sewers with economy and advantage through quick-sands, or where the strata is full of water ; also in narrow streets where deep trenches have to be excavated. A cast-iron sewer may be two-thirds the diameter of an earthenware pipe or a brick sewer, as the cast-iron pipe may work full and even under pressure.

Main sewers require to be provided with storm overflows in case of heavy rains.

Sewers should not join at right angles, unless the curve and extra fall is provided in the manhole.

Tributary sewers should deliver sewage in the direction of the mainflow.   Fig. 63.

Sewers and drains at junctions and curves should have extra fall to compensate for friction.

Sewers of unequal sectional diameters should not join with level inverts, but the lesser, or tributary sewer, should have a fall into the main sewer at least equal to the difference in the sectional diameter.   If the inverts of tributary sewers are not above, or, at least, are not up to the level of the ordinary flow of sewage in the main sewer, such tributary sewers, or drains, will be liable to be backwatered, in which case deposit will take place in the length of submerged invert, and so the tributary sewer, or drain, will become choked with its own silt.

Pipe drains should have the joints made with cement.   In bad soils the joints should also be bedded in concrete, a gap being cut to receive the concrete so that the pipes may bed evenly.

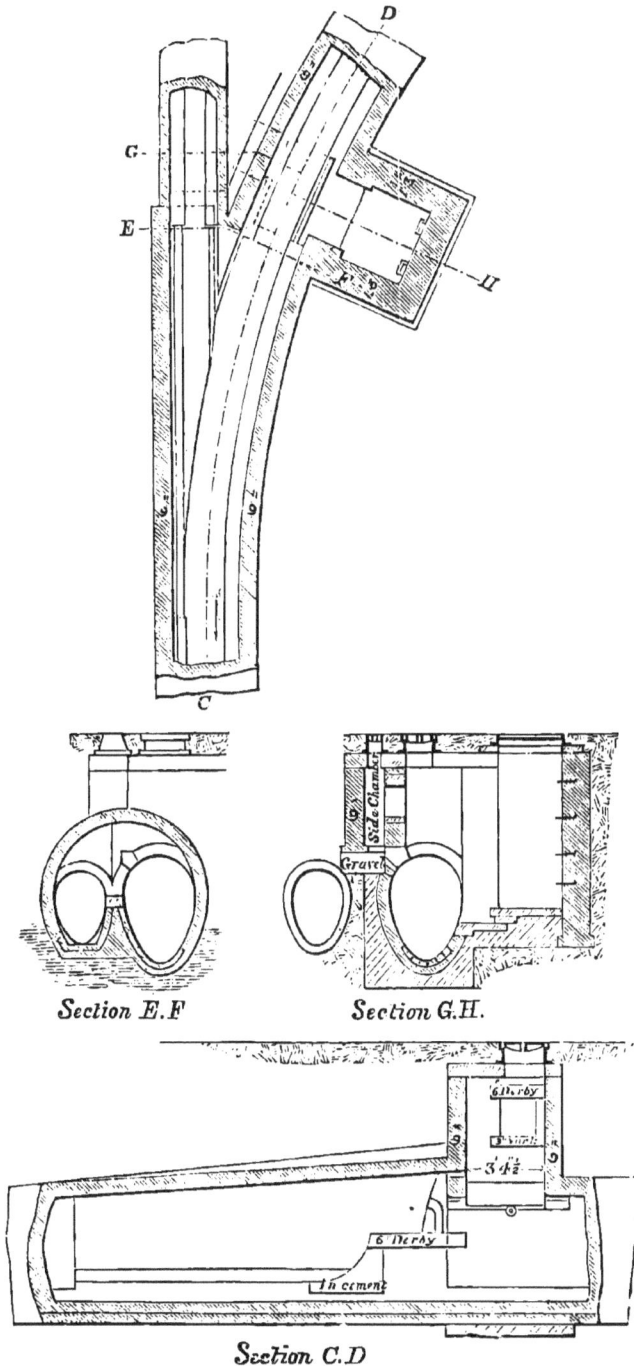

Section E.F   Section G.H.

Section C.D

Fig. 63. Sewer-junction 3′ 0″ × 2′ 0″ into 4′ 6″ × 3′ 0″ showing side entrance.

T

Sewer pipes of more than eighteen inches diameter are troublesome to handle and difficult to joint, and do not make such good work as a brick drain.

Earthenware pipes of equal diameter should not be laid as tributaries, but a lesser pipe should be joined on to a greater.

Side openings for house drains should be provided in the original construction of all sewers wherever it is probable they will be required. Every effort should be made by careful construction to prevent sewage and sewer-gas from passing out of the sewer into the subsoil. Where subsoil drainage is necessary in connexion with sewers, either the subsoil water should be carried off by separate drains or the mouth of the subsoil drain should terminate in a ventilated disconnecting manhole, whence it should pass into the sewer.

Flushing Sewers.

Flushing a sewer means accumulating water at some point in the sewer in sufficient quantity to pass down and along the portion of the sewer below with a rush when suddenly liberated, in order that the water may loosen and carry away all sediment.

In a perfect system of sewers, flushing should be unnecessary: but perfection in a sewer depends as much on the excellence of workmanship as on the design ; and any imperfections lead to deposit and its corresponding evils.

Every care should be taken to prevent heavy substances entering the sewers by properly formed gullies to catch and retain the material ; especially where road washings pass into the sewers; and it may in some exceptional cases be desirable to form catch-tanks in the sewer. These catch-tanks should be wider than the sewer, so as to cause a check in the flow, and thus cause heavy materials to be deposited, and the bottom should be below the invert of the sewer so as to retain the materials when dropped. The materials thus dropped should be dredged out, at intervals, as found necessary.

As a general rule, however, sewers when true in line and gradient, and sound in construction, and provided with side-entrances, manholes, and flushing-chambers, at proper intervals, will permit of silt being flushed through to the outlet.

In a system of sewers every manhole may be a flushing chamber, provided with flap or flushing boards so as to be charged with water for flushing purposes. There should also be a flushing chamber at the head of each sewer and drain, and every flushing chamber should be permanently ventilated.

Mr. Roger Field's flush tank, with his automatic syphon already described, affords a convenient means of periodically flushing drains or moderate-sized sewers by the flow of water taps or from the overflow of a drinking fountain, or other small continuous flow which is generally wasted.

The volume of water necessary for flushing necessarily varies with the diameter of the sewer or drain and its inclination. A flush for a 4-inch w. c. soilpipe has recently been fixed at 3 gallons, but in any case it should not be less than 2 gallons. For the adequate flush of a 4-inch drain-pipe 17 gallons are necessary ; a 6-inch drain requires 44 gallons, a 9-inch requires 100 gallons, a 15-inch drain requires 270 gallons, and a 3-foot sewer requires 1800 gallons. The distance to which the effects of the flush is felt depends on the inclination, and therefore in flushing a line of sewer the intervals at which the flush should be repeated depend on the character of the sewer.

It is possible to injure sewers by overflushing them.

In towns where there is a regular water supply the flushing chambers may be filled by the flow in the drains, otherwise it may be necessary to resort to water carts.

Springs of water, and the water from canals, reservoirs, rivers, and streams, may occasionally be so situated as to be easily made available for purposes of sewer-flushing.

In the London sewers the flushing is effected by flushing-

gates provided at certain points. In the Paris and Brussels
sewers a moveable flushing apparatus is provided.

The sewer has a footpath on each side, along which a truck
is arranged to run as on a railway, as shown in Fig. 64. The
truck is provided with a board of the exact section of the

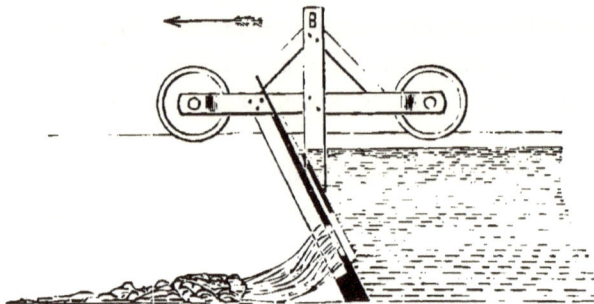

Fig. 64. Moveable flushing apparatus.

sewer, which can be raised or lowered by a screw apparatus
fitted to the truck. When the board is lowered, the water in
the sewer is dammed back. Two small doors are provided in
the board near the bottom, and by opening these when the
sewage is dammed back, a strong current of water is passed

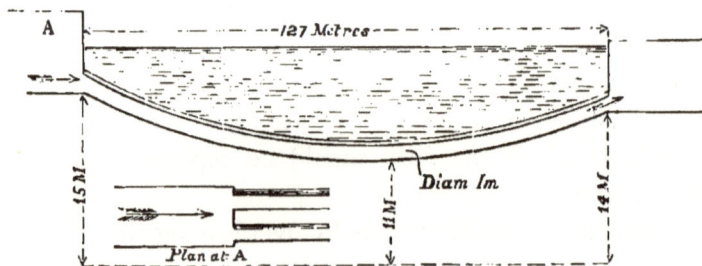

Fig. 65. Sewage syphons under the Seine at Paris.

through so as to wash away any accumulation of débris. The
truck can be moved on to any spot where the action is required.

The sewers at Paris are carried under the Seine from one
bank to another by means of pipes laid in the form of an
inverted syphon. (Fig. 65.) Each pipe is about one metre in

diameter ; to clean out the pipes a wooden ball of a diameter a little less than that of the tube is periodically caused to roll through the pipe. It delays the flow, and thus accumulates a head of water behind it, sufficient to force away any impediment from the bottom of the syphon.

There cannot be any strict rule by which to ascertain the weight of mud any special town sewage will deposit during a day, week, month, or year, as the weather necessarily affects the condition of road and street surfaces, and it is (for the most part) from dirty yards and street-surfaces that sewage-mud is derived. A wet season will, as a rule, give from any area more mud than a dry season. The relative weight and character of the mud will also be determined by the character of the street-surfaces, and the weight and rapidity of the traffic. Paved streets produce less mud than macadamised roads, whilst good scavenging prevents an excess of mud being washed from the street surfaces into and through the drains and sewers.

With respect to wear of streets and roads, rapid traffic is destructive, heavy and rapid traffic being most destructive. A good foundation for a road or street is of the first importance, because much of the mud found on pavements has worked up from below. Consequently all street surfaces should be laid on a bed of concrete : by this means not only will surface mud be diminished but the road will last longer ; a well-formed road, though more costly to make, will be more cheaply maintained. It is also of great importance to use all proper means to prevent grit and mud being washed into drains and sewers. For this purpose all street gullies, instead of opening directly into the sewers, should be provided with catch-pits constructed to catch and retain the detritus from the street washings, and frequently cleaned out.

CHAPTER XIX.

DISPOSAL OF WATER-CARRIED SEWAGE.

THE Rivers Pollution Prevention Act, 1876, forbids the pollution of rivers and streams by the admission of crude sewage.

At first sight, the simplest mode of getting rid of sewage is to place it in the sea.

There are, however, towns both on the sea-shore and on tidal estuaries in which a nuisance is caused by the discharge of crude sewage too near to roads, houses, or bathing-places.

Moreover there are many places where this method of disposing of sewage is inapplicable; it would be wasteful except for the addition to fish life to which its presence gives rise.

There is, however, no doubt that water carriage for the removal of excreta is the cheapest and most convenient system, so far as the transport to a distance is itself concerned. It is only when the sewage thus removed has to be finally disposed of without creating a nuisance that the difficulties arise.

To dispose of sewage therefore either the outlets must be extended into the sea, and the sewage be so poured in that it shall not be returned on to the shore: or else it must be pumped inland for irrigation, or the solids must be precipitated, and the subsequent effluent clarified by some mode of filtration so that it shall attain to the condition required by the Rivers Pollution Act before it can be allowed to flow into a stream or river.

The degree of purity to be required in clarified sewage before it is admitted into a river, depends to some extent on the degree of purity of the river water into which it is to flow. For instance, a different standard of purity might reasonably be admitted for water turned into the Thames at Gravesend from that required at Oxford or Reading.

The cost of irrigation or disinfection and clarifying sewage so as to prepare it for passing into a stream must be regarded as expenditure of money for cleansing the town to whose sewage it is applied.

The disposal of sewage is a question on which the Sanitary Engineer has expended a great share of attention, without attaining a result which all parties have been willing to accept as a final settlement of that question. One reason of this arises from the fact that the question of *utilisation* of sewage has been mixed up with that of *removal* of sewage.

The desire to make a profit has, in many cases, become paramount, and people have been unwilling to recognise that whilst theory shows that the utilisation of sewage in manuring the land should repay the cost, in practice there are various barriers to financial success. Moreover, the population in different localities is living under very various conditions. In one part of the country the population is scattered—and plenty of land unoccupied by dwellings can be found for sewage irrigation. In other parts the population is so dense that there is little room for sewage farms at any reasonable distance from towns. In some towns chemical products have been allowed to find their way into sewers, and this has rendered the sewage unfit to apply to land.

The inventors of many of the systems for clarifying sewage-water have generally advocated their projects on the ground that profit will be derived from the manure which is to be extracted from the sewage. If there had been no idea of profit, the numbers of inventors would have been more limited, and the subject would have received less consideration ; and therefore. looking at it from this point of view, it has been advantageous for science that the idea of profit has so largely prevailed.

On the other hand, the various corporations stayed their hands for many years in providing drainage and water supply, in order to wait for the discovery of this modern philosopher's stone, and thus much sickness, death, and misery have continued in the country, which more energetic measures would have prevented.

It is no doubt true that sewage contains matter of enormous value. The value of the manurial constituents of the sewage of London was estimated by Dr. Hofmann in 1857 at nearly £1,500,000 per annum ; but he added :—'An enquiry into the nature of the valuable matter carried off in our sewers, an attentive examination of the chemical properties of the constituents, together with the consideration of the extraordinary and constantly-increasing degree of dilution in which they exist, cannot fail to impress the chemist, on purely theoretical grounds, with the magnitude of the difficulties which oppose themselves to the successful accomplishment of the task.'

Nor do these difficulties diminish, if the question be submitted to the test of experiment.

The valuable constituents of sewage are like the gold in the sand of the Rhine; its aggregate value must be immense. but no company has yet succeeded in raising the treasure.

The character of the sewage of different towns varies with the industrial pursuits in those towns ; in towns. however, where no special trades exist, the sewage is of a uniform character.

The limits of this chapter merely permit of a brief allusion to some of the processes which have been proposed.

The schemes for defecating sewage are based on precipitating the solid constituents. The object of precipitation is to remove in a solid, dry, or semi-dry state the putrescible constituents of the sewage, and to render the filtrate or effluent water sufficiently pure to mingle with streams, or to be employed for purposes of irrigation.

There are two points connected with the effluent from precipitation processes which ought to be borne in mind. In the first place an alkaline effluent must be avoided, because wherever there is an alkalinity there is a tendency to putrefaction. Where the effluent can be kept acid it is safe. Another point, urged many years ago by Professor Heisch, is that there must not be phosphoric acid in the effluent, or there will be a tendency to produce the low confervoid growth commonly called sewage fungus. That is a difficulty with any process which employs phosphates.

The lime process is the simplest, and was the first to attain prominence. It consists of mixing cream of lime with sewage in the proportion of about sixteen grains of lime to the gallon. The mixture is then allowed to subside, and the clear water flows off. With recent sewage, and on a small scale, the deodorisation is tolerably complete, but it leaves in the effluent water at least four-fifths of the soluble organic matter of the sewage. A mixture of lime and sulphate of alumina is capable of removing a larger quantity of matter from sewage than lime alone, and much more quickly. This process not only removes the whole of the suspended matter, but also a considerable quantity of dissolved matter, both mineral and organic; while with the lime process a minute quantity even of suspended mineral matter is left unremoved.

The A B C process consists of using alum, blood, clay, and charcoal. The clay and charcoal, and where necessary a little lime, are finely ground up with water to form an emulsion, and mixed with the sewage; a solution of sulphate of alumina is then added.

The sewage being a slightly alkaline liquid charged with nitrogenous organic matter, the sulphate of alumina is decomposed; the alumina is separated in a flocculent state, and, by virtue of its affinity for dissolved organic matter, when it sinks it drags down with it a corresponding amount of nitrogenous impurity : the blood is a liquid charged with albumen ; albumen is coagulated in the presence of alum, and in the same way as this coagulability of albumen is utilised in fining wine and coffee, so it is made use of in this process to join with the alumina in its precipitation, and to assist in the further removal of the putrescible constituents out of the solution.

The immediate action of the sulphate of alumina depends on the presence of natural alkali in the sewage produced by the decomposition of urea in carbonate of ammonia ; but a great deal of the sulphate of alumina is sometimes wasted because the sewage is not sufficiently alkaline to precipitate the whole of it : when this is the case lime must be added.

The effluent water is stated to be sufficiently purified for admission into a running stream without being a nuisance or injurious to public health ; but the question remains as to how far the cost of the process can be covered by the value of the residuum.

The phosphate of alumina process consists of employing prepared phosphates, the action of which is to curdle, and separate the fœcal matter in the sewage, and to add lime to separate the soluble phosphates. The solid matter is collected in precipitation tanks. The effluent is said to be sufficiently pure for admission into a river of the purity of the lower part of the Thames, where the water is not used for drinking purposes.

The system of the General Sewage and Manure Company, or sulphate of alumina process, is first to pass the sewage through strainers, the solid matter thus retained is applied directly as manure ; the strained sewage is mixed with a solution of sulphate of alumina; after which it receives an addition of cream of lime, the mixture being thoroughly agitated after the addition of each chemical.

It is then passed into precipitation tanks, of which it passes through three in succession. The effluent water flows in a thin sheet from the surface to filtering beds, from which it percolates through a depth of five feet of earth and then passes into the river in a clear bright condition.

The filter beds are used intermittently and are planted with osiers and rye grass.

The sludge or deposit recovered from the precipitating tanks has to be dried.

In all these systems of precipitation a large portion of the fertilising matter passes off in the effluent water, and the disposal of the sludge is a great difficulty. In proportion as the quantity of the precipitating material is increased, the value of the sludge or manure is diminished. It must be dried, and unless heat is applied, the space required for its stowage would be enormous. General Scott proposed to utilise the organic matter in the sludge as fuel, and to produce lime compounds to be used as cement, or for agricultural purposes. He adds to the sewage a sufficient quantity of slaked lime to produce a complete precipitation of the acids present in sewage, and enough clay to form, with the silica and alumina already present in the sewer water, about 20 per cent. of the bulk of the calcined sewage sludge. When the sludge is dry it is introduced into a kiln with a small quantity of fuel to commence the combustion, the organic matter in the sludge furnishes the rest of the fuel required. A million gallons of ordinary town sewage treated with $1\frac{1}{4}$ tons of slaked lime yield 2 tons of Portland cement.

This brief account of a few of the methods which have been adopted for clarifying sewage, sufficiently illustrate this branch of the subject.

Dr. Frankland stated that the precipitation processes which have been proposed for the purification of sewage, are not sufficiently effective for rendering the water which has been contaminated fit for domestic purposes. But all these processes more or less purify the sewage, and render it more or less fit for turning into a river, or for further purification by application to land.

The fertilising matter in sewage can however be effectually removed and utilised, and the effluent water can be rendered sufficiently pure to flow into streams by passing the fresh crude sewage through the soil by irrigation in thin films, under proper and favourable conditions of land covered with growing crops. Purification requires the co-operation of living organisms who convert the organic soluble matter into nitrates and nitrites. The loam on the surface at once supplies these nitrifying organisms, which are ready to convert the sewage into a form suited for food for plants growing on the land. By this means the trouble, nuisance and cost of precipitating, screening and filtering are avoided, and the whole of the fresh sewage, sludge and fluid, is incorporated with the land.

The difficulty of sewage irrigation lies in the impossibility of always securing the necessary favourable conditions ; and it may become necessary in some cases to combine precipitation, involving removal of the sludge with the flow of the cleared effluent over land.

The best land for a sewage-farm would have a free loamy soil, and open subsoil, with a sufficient proportion of clay to moderate the percolative powers of its other constituents ; the surface tolerably even, having a southern aspect, and gently sloping to the south.

Where clay soils must be resorted to, the drains must be frequent, in order to overcome the natural retentiveness of the land ; and they should be so laid out as to allow the sewage if possible to be applied twice—that is to say, the under-drains of the land to which the sewage is first applied should discharge their effluent upon a lower surface of land, by which any impurity retained after the first application may be removed by the second.

Clay land requires deep draining, and the surface should be well broken up, either by spade labour or by deep steam-ploughing. The surface should be deeply and perfectly trenched, so as to secure that disintegration of the soil which will prevent its cracking, and give it a uniform filtering power ; the drains must be so laid and protected as to remove subsoil-water, after filtration through the soil ; they should not allow either unfiltered surface-water or sewage to pass through cracks direct to the drains.

Clay soil, properly subsoiled and under-drained, would produce better crops than light sandy soils. The reason is simple. In clay soils there are found in great abundance all those mineral matters which plants take up as food ; whereas, there is a deficiency of mineral plant-food in light soils, and the excess of nitrogenous matters in sewage produces over-luxuriance of growth in the plants grown on light sandy soils. In clay soils the excess of mineral matters counter-balances the excess of nitrogenous compounds in sewage, and more healthy and heavier crops can be obtained, by means of sewage irrigation, on well-drained and well-worked clay soils than on light sandy soils.

Under-drainage and surface preparation are necessary, the one to prevent wetness and the other to secure an even distribution of the sewage over the surface, and each must be regulated by the character of the soil.

Drainage can never do harm, while if the subsoil partakes of a retentive character drainage is indispensable. In soils comparatively free in character but few drains are required.

The inclination which it is necessary to give to the surface of land prepared for irrigation will entirely depend upon the character of the soil and subsoil. The area of a sewage-farm must be laid out in slopes, regulated in size and position by the configuration of the surface and the degree of porosity the soil possesses. With very free soils, a gradient of 1 in 25, or as much as 1 in 20, may be necessary to gain an overflow which will cover the surface and prevent excessive absorption, whilst on very suitable soils a flat slope of 1 in 150 may be a workable gradient; the sewage-carriers should be shallow trenches carried along the ridge nearly level, and made to overflow evenly over the surface by damming up at intervals. Whenever their position is likely to be permanent, they should be made of bricks, stone, or concrete.

Every area, however rough or uneven, may have level contour lines set out over its entire surface, so that by forming conduits on these contour lines the surface may be irrigated.

Land having an irregular and steeply sloping surface may have sewage-intercepting drains and carriers so arranged as to intercept the sewage from the upper areas and bring it over the lower areas a second or a third time, by such means more effectually purifying the sewage.

Roads over a sewage farm should, where practicable, be along the fences; costly permanent roads are not required. When land has been properly prepared for the reception of sewage, it may be irrigated in all weathers, so as to purify the sewage.

A wet season does not necessarily injure a sewage farm if the means of removing and consuming the produce are equal to the growth of the crops.

One gallon of sewage weighs 10 lbs.; 224 gallons, or 2240 lbs. are one ton; 22,400 gallons, or 224,000 lbs. are 100 tons —equal to nearly one inch in depth over one acre of land.

A dressing of liquid sewage half an inch deep amounts to nearly 50 tons per acre, a quantity more than sufficient to fertilise any growing crop.

Ten inches equals 1000 tons, and 12,000 tons per acre per annum equals 120 inches in depth, and this volume may be used on well-prepared land without swamping it, as land will filter several inches in depth per day, when the sewage is equally and evenly distributed. Looked at from a productive point of view, this is a wasteful application of sewage, but it answers as a means of purification.

Italian rye grass will dispose of a larger quantity of sewage than almost any crop, and give heavy crops if the roots are young. The greatest producing power will be in the first year's growth. A second year is probably the utmost length of time it should be in the ground.

No larger area of Italian rye grass should be sown than will admit of the grass upon it being disposed of in the district, as it will not keep, nor will it bear distant carriage. Sewage-grown grass will however make good and wholesome hay if the season will permit, or if the grass can be artificially dried.

To give a sewage farm the chance of paying, the land must be obtained at a reasonable price, and the cost of preparation must be moderate; there must also be reasonable skill in cropping, in cultivation, and in management; under such conditions land irrigated with sewage ought to pay a reasonable rent. If steam power has to be used for pumping the sewage, this of course must be paid for in addition.

Sewage has been valued as a manure at from $\frac{1}{2}d.$ to $2d.$ per ton. The same sewage will, however, be worth $2d.$ in a dry summer, which may not be worth even a halfpenny to the farmer in a wet season and through the winter.

In estimating the quantity of land to be allotted to a sewage-farm it would be advisable as a rule to take one acre for 100 persons.

The sewage has to be applied carefully, so as not to injure the growing plants; it cannot be continually applied, as the matter in suspension has a tendency to choke the pores of the soil. It must be applied at proper times; and when not applied at once the suspended matter must be removed by precipitation or filtration. Moreover, it should not be applied in proximity to dwellings.

In places where land free from building is difficult to be obtained, or where the amount of land available is not sufficient for broad irrigation, those conditions which are acting in the soil to convert the sewage into food for plant life, may be utilized in a concentrated form by means of intermittent filtration. In this case it was originally held that one acre should be alloted to 1000 persons; but the knowledge which has been acquired of the methods by which purification is effected has enormously added to the efficiency of intermittent filtration.

Whilst irrigation may be described as the distribution of the sewage without supersaturation of the land, having in view the production of a maximum growth of vegetation, intermittent filtration is the concentration of the sewage at short intervals on as few acres of land as will absorb and cleanse it, by calling in the co-operation of living nitrifying organisms.

The simplest theory of the working of any filter is that it is a more or less perfect strainer. In this aspect the working of the filter is continuous, but it soon chokes and must be cleaned.

The intermittent filter, on the other hand, is no longer a mere mechanical strainer. When first established there may be a period at the outset when it effects little more than a mechanical purification ; but, under proper conditions, the filter becomes a method of developing the action of bacteria by the exposure of the sewage in the presence of air.

The Massachusetts experiments on the purification of sewage show that a sand filter does not effect the nitrification when first used. Time is necessary for it to accumulate a suitable colony of bacteria. Furthermore, the colony adjusts itself to the work it has to do. If, then, the amount of sewage is suddenly increased, and is continued at the larger amount, the nitrification will at first be incomplete, but the bacteria will soon multiply and purification will again become satisfactory, often amounting to the destruction of 99½ per cent. of the nitrogenous matters in the sewage, and all but a fraction of one per cent. of the bacteria.

Nitrification is affected by the season and by temperature. It is most active in the growing months of May and June—even more so than in the hotter months of July and August. With this exception the amount of nitrification varies with the amount of the sum of the ammonias in the sewage, so that, in the winter months of 1888-89, while the nitrates of the effluent were lower than at other times, it was found that the sum of the ammonias in the sewage was also lower, and that nitrification at that time was quite as complete as in the previous months.

The general conclusions were thus summed up in the report of the chemist to the experiments, Mr. Hazen:—

'The purification of sewage by intermittent filtration depends upon oxygen and time; all other conditions are secondary. Temperature has only a minor influence; the organisms necessary for purification are sure to establish themselves in a filter before it has been long in use. Imperfect purification for any considerable period, can invariably be traced either to a lack of oxygen in the pores of the filter, or to the sewage passing so quickly through, that there is not sufficient time for the oxidation processes to take place. Any treatment which keeps all particles of sewage distributed over the surface of sand particles, in contact with an excess of air, for a sufficient time, is sure to give a well-oxidized effluent; and the power of any material to purify sewage depends almost entirely upon its ability to hold the sewage in contact with air. *It must hold both air and sewage in sufficient amounts.'*

Mr. Lowcock, C.E., has proposed a filter in which, by the continued supply of air into the body of the filter, the necessary means of subsistence would be afforded to the nitrifying organisms to enable them to carry on the purification of the sewage. Colonel Waring, in the United States, proposed an analogous method; but he adopted the plan of draining the filter every hour and of vigorously aerating it for, say, five minutes, thoroughly changing the air in all its voids, thus storing sufficient oxygen to keep the bacteria active until the

U

next period of aeration. He further suggested that the filtering material should be divided into sections to be aerated vigorously in turn, by which means the entire filter could be satisfactorily aerated with one-half of the power and air which would be required if the aeration were constant.

In Mr. Lowcock's experiments the quantity treated as the most efficient rate is stated to have nearly equalled the sewage from 17,000 persons per acre. Colonel Waring states that the maximum flow through the aerators after nitrification began was at the rate of 5,000,000 gallons per acre.

Present Aspect of the Sewage Question.

The following are the general conclusions to which a consideration of the several methods of disposing of sewage leads :

1. No one system of sewage disposal could be adopted universally : the peculiarities of different localities require different methods.

2. With dry systems, where collection at short intervals is properly carried out, the result, as regards health, appears to be satisfactory.

3. By some of the various processes, based upon subsidence, precipitation, or filtration, a sufficiently purified effluent can be produced for discharge, without injurious result into watercourses and rivers, provided they are of sufficient magnitude to effect a considerable dilution. In the case of towns, where land is not readily obtained at a moderate price, those particular processes afford suitable means of disposing of water-carried sewage ; but the sludge in a manurial point of view is of low and uncertain commercial value ; hence means must be found for getting rid of it without reference to possible profit.

4. The Massachusetts experiments show that :

(*a*) The suspended matters of sewers (sludge) can be mechanically withheld by straining slowly through suitable material.

(*b*) The filth accumulated by this straining material can be destroyed, and the straining medium restored to a clean condition by mere aeration.

(*c*) The successive alternate operations of fouling and cleansing can be carried on indefinitely without renewal of the straining material.

(*d*) The purification obtained by this straining process practically equals that accomplished by chemical precipitation, and is sufficient to admit of discharge into any considerable body of water not used as a source of domestic supply, or for manufacturing purposes requiring great purity.

(*e*) Such filters can be maintained in *constant* and *efficient* operation by suitable aeration.

(*f*) The erection of a plant capable of purifying large volumes of sewage upon a relatively small area calls for no costly construction. Repairs and renewals are merely nominal. The attendance required is but slight. There is no outlay for chemicals, &c. The only expense of mechanical operation is the driving of the blower or air-compressor.

(*g*) The process admits of wide variation in the selection of filtering material, and nearly every community can find, in its local resources, something suitable for the purpose.

5. In localities where land at a reasonable price can be procured, with favourable natural gradients, with soil of a suitable quality, and in a sufficient quantity, a sewage farm, if properly conducted, is apparently the best method of disposing of water-carried sewage. It is essential, however, to bear in mind that a profit should not be looked for by the locality establishing the sewage farm, and only a moderate profit by the farmer.

6. As a rule no profit can be derived at present from sewage utilisation.

7. For health's sake, without consideration of commercial profit, sewage and excreta must be got rid of at any cost.

U 2

CHAPTER XX.

CONCLUSION.

THE conditions which should govern the healthy construction of dwellings are embodied in pure air and pure water. Five hundred years ago the population of the kingdom was only equal to the present population of the metropolis. When the first census was taken in 1801, the population of England and Wales was less than 9,000,000; it has now reached 30,000,000. We are crowded together as we were never crowded before; our pursuits are more sedentary, our habits more luxurious; houses increase in number; land is more valuable, the green fields more remote; our children are reared among bricks and paving-stones; the public health can only be maintained by special sanitary appliances and precautions.

It has therefore become impossible in the question of health for any one member of a community to separate his interest from that of his neighbours. If he places his house away from others, the air which he breathes may receive contamination from the neighbouring district; the dirty water which he throws away may pollute the stream from which his neighbours draw their supply; and when a population congregates into towns, the influence of the proceedings of each individual on his neighbour becomes strongly apparent.

In places where many dwellings are congregated together, the requirements for health may be classed under the heads, first, of those that are common to the community, such as the supply of good water, the removal of foul water, and the removal of refuse matter; and secondly, of those which immediately concern the individual householder, such as the condition of his house and the circumstances of its occupation.

But the existence of some danger to health in a house in a town or village may be a source of danger to the houses around. Thus it is the interest of every person in a community of houses, that every other member of the community should live under conditions favourable to health.

Each year, as civilization and population increase, so do these considerations increase in importance. So long as preventible disease exists in this country, we must not delude ourselves with the idea that we have done more than touch the borders of sanitary improvement.

Laws alone can do little to remove or prevent sanitary defects. A central department of the government may assist in spreading through the community the knowledge of what is necessary, and may publish practical methods of applying that knowledge. To that extent government may usefully interfere. But real and permanent sanitary progress can only be obtained by the efforts of the people themselves; by the education of the nation in a knowledge of the laws of health; and by the creation of an efficient local administration endowed with adequate power and responsibility.

The first step therefore towards further progress is to imbue the owners and occupiers of houses and cottages with a knowledge of the laws of health: they are the laws of common sense: simple and economical methods of applying them are generally those found most effectual.

The health, intelligence, morality, and general well-being of a community depends upon the condition of the dwellings.

The local authorities in towns and villages may direct unhealthy dwellings to be pulled down, but the removal of insanitary dwellings should not be accompanied by over-crowding in other dwellings. When therefore in such cases private enterprise fails to supply new houses, it should be the duty of the local authorities to build healthy dwellings out of the rates to replace the dwellings so removed.

A supply of pure drinking water within the reach of all the inhabitants should be provided ; as well as such a degree of drainage as the local conditions show to be necessary for ensuring that all fouled water would be removed rapidly, and not be allowed to stagnate in ditches or on the surface, nor to pass into streams until it has been clarified by passing over land or otherwise.

All refuse should be rapidly removed from the immediate vicinity of dwellings.

The plans of all new dwellings and important alterations of existing dwellings over the whole country should be subject to a general Building Act containing provisions based on the principles suggested in this volume, to be enforced by the local authorities ; and whenever the local medical officer has reason to suspect that a cause of disease exists in any house in a town or village, there should be a power to enter and inspect the premises, and to require the removal, at the expense of the owner or occupier, of any cause of disease found to exist. Such arrangements would require that all congregations of houses in the country districts should be subjected to the jurisdiction of a local surveyor, such as is the case generally in towns.

It is the function of the sanitary engineer or local surveyor to adopt measures to prevent or to remove those sources of danger to health which the medical officer is called upon to detect.

The community does not permit any man to practise medicine without having satisfied a careful and responsible board of examiners that he has educated himself for his position: and education in the principles of sanitary science is just as necessary to ensure the efficient fulfilment of the duties of a sanitary engineer or local surveyor, as is the study of medicine to the medical man. Sanitary science is somewhat new, it rests on the attentive observation of facts; its true principles have been slowly and painfully collected during a long period, from a careful observation of human disease and misery, by those who have preferred the study of facts to the more enticing but more fallacious creation of theories; hence sanitary science is built up from details; wherever the details have been carefully and intelligently applied success has invariably followed their application. When the public realise that the progress of the nation in healthiness is to be attained by a careful attention to these details, they will insist that the local surveyor and sanitary engineer shall have a complete education in the science of the healthy construction of buildings, and in the arrangements for health to be adopted in towns and villages; that is to say, in the conditions necessary for the prevention of disease; just as at the present time they require education in those who minister to the cure of disease.

The various processes which are necessarily incidental to life, especially to life in crowded communities, contribute largely to that deterioration of air and water which is a principal cause of preventible disease. But the operation of a free atmosphere and of running streams, provides a ready means of purifying the air and water thus contaminated.

The evils which have arisen from this deterioration have been gigantic, in consequence of the apathy of the community: an apathy which results from an ignorance of the cause of these evils, and of the means of remedying them. But the remedy for these evils would be comparatively easy if each member of the community were induced to perform his part in their prevention or removal. In order to attain

this end, every member of the community should be taught
the principles of sanitary knowledge. When this has been
done, and when the co-operation of every individual in
a community has been enlisted to aid in enforcing attention
to sanitary details, we may hope for practical progress in
the diminution of preventible disease, and for a general
improvement in the health, and therefore in the happiness,
as well as in the wealth-producing power, of the community.

INDEX.

A.

THE END.

BY THE SAME AUTHOR

———•———

HEALTHY HOSPITALS

OBSERVATIONS
ON SOME POINTS CONNECTED WITH
HOSPITAL CONSTRUCTION

BY

SIR DOUGLAS GALTON

WITH ILLUSTRATIONS

Oxford
AT THE CLARENDON PRESS
LONDON: HENRY FROWDE
OXFORD UNIVERSITY PRESS WAREHOUSE, AMEN CORNER, E.C.

20, 8/00

Clarendon Press, Oxford.

SELECT LIST OF STANDARD WORKS.

1. DICTIONARIES.

A NEW ENGLISH DICTIONARY

ON HISTORICAL PRINCIPLES,

Founded mainly on the materials collected by the Philological Society.

Imperial 4to.

EDITED BY DR. MURRAY.

PRESENT STATE OF THE WORK.

			£	s.	d.
Vol. I. { A / B } By Dr. MURRAY . . . . Half-morocco			2	12	6
Vol. II. C By Dr. MURRAY . . . . Half-morocco			2	12	6
Vol. III. { D } By Dr. MURRAY / { E } By Mr. HENRY BRADLEY } . . . Half-morocco			2	12	6
Vol. IV. { F } By Mr. HENRY / { G } BRADLEY	F-Field		0	7	6
	Field-Frankish		0	12	6
	Franklaw-Glass-cloth . . .		0	12	6
	Glass-coach-Graded		0	5	0
	Gradely-Greement		0	2	6
Vol. V. H—K By Dr. MURRAY.	H-Hod		0	12	6
	Hod-Horizontal		0	2	6
	Horizontality-Hywe . . .		0	5	0
	I-In		0	5	0
	In-Infer		0	5	0
	Inferable-Inpushing . . .		0	2	6

☞ *The remainder of the work, to the end of the alphabet, is in an advanced state of preparation.*

*** *The Dictionary is also, as heretofore, issued in the original Parts—*

					£	s.	d.
Series I.	Parts I-IX.	A—Distrustful	each	0	12	6
Series I.	Part X.	Distrustfully—Dziggetai		0	7	6
Series II.	Parts I-IV.	E—Glass-cloth	each	0	12	6
Series III.	Part I.	H—Hod		0	12	6
Series III.	Part II.	Hod—Hywe		0	7	6
Series III.	Part III.	I—Inpushing		0	12	6

Oxford: Clarendon Press. London: HENRY FROWDE, Amen Corner, E.C.

A Hebrew and English Lexicon of the Old Testament, with an Appendix containing the Biblical Aramaic, based on the Thesaurus and Lexicon of Gesenius, by Francis Brown, D.D., S. R. Driver, D.D., and C. A. Briggs, D.D. Parts I–VIII. Small 4to, 2s. 6d. each.

Thesaurus Syriacus : collegerunt Quatremère, Bernstein, Lorsbach, Arnoldi, Agrell, Field, Roediger: edidit R. Payne Smith, S.T.P. Vol. I, containing Fasciculi I–V, sm. fol., 5l. 5s.
₊ *The First Five Fasciculi may also be had separately.*
Fasc. VI. 1l. 1s.; VII. 1l. 11s. 6d.; VIII. 1l. 16s.; IX. 1l. 5s.; X. Pars. I. 1l. 16s.

A Compendious Syriac Dictionary, founded upon the above. Edited by Mrs. Margoliouth. Parts I and II. Small 4to, 8s. 6d. net each.
₊ *The Work will be completed in Four Parts.*

A Sanskrit-English Dictionary. Etymologically and Philologically arranged, with special reference to cognate Indo-European Languages. By Sir M. Monier-Williams, M.A., K.C.I.E.; with the collaboration of Prof. E. Leumann, Ph.D.; Prof. C. Cappeller, Ph.D.; and other scholars. *New Edition, greatly Enlarged and Improved.* Cloth, bevelled edges, 3l. 13s. 6d.; half-morocco, 4l. 4s.

A Greek-English Lexicon. By H. G. Liddell, D.D., and Robert Scott, D.D. *Eighth Edition, Revised.* 4to. 1l. 16s.

An Etymological Dictionary of the English Language, arranged on an Historical Basis. By W. W. Skeat, Litt.D. *Third Edition.* 4to. 2l. 4s.

A Middle-English Dictionary. By F. H. Stratmann. A new edition, by H. Bradley, M.A. 4to, half-morocco, 1l. 11s. 6d.

The Student's Dictionary of Anglo-Saxon. By H. Sweet, M.A., Ph.D., LL.D. Small 4to, 8s. 6d. net.

An Anglo-Saxon Dictionary, based on the MS. collections of the late Joseph Bosworth, D.D. Edited and enlarged by Prof. T. N. Toller, M.A. Parts I–III. A–SÁR. 4to, stiff covers, 15s. each. Part IV, § 1, SÁR–SWÍÐRIAN. Stiff covers, 8s. 6d. Part IV, § 2, SWÍÞ-SNEL-ÝTMEST, 18s. 6d.
₊ *A Supplement, which will complete the Work, is in active preparation.*

An Icelandic-English Dictionary, based on the MS. collections of the late Richard Cleasby. Enlarged and completed by G. Vigfússon, M.A. 4to. 3l. 7s.

2. LAW.

Anson. *Principles of the English Law of Contract, and of Agency in its Relation to Contract.* By Sir W. R. Anson, D.C.L. *Ninth Edition.* 8vo. 10s. 6d.

—— *Law and Custom of the Constitution.* 2 vols. 8vo.
Part I. Parliament. *Third Edition.* 12s. 6d.
Part II. The Crown. *Second Edition.* 14s.

Baden-Powell. *Land-Systems of British India;* being a Manual of the Land-Tenures, and of the Systems of Land-Revenue Administration prevalent in the several Provinces. By B. H. Baden-Powell, C.I.E. 3 vols. 8vo. 3l. 3s.

Digby. *An Introduction to the History of the Law of Real Property.* By Sir Kenelm E. Digby, M.A. *Fifth Edition.* 8vo. 12s. 6d.

Grueber. *Lex Aquilia.* By Erwin Grueber, Dr. Jur., M.A. 8vo. 10s. 6d.

Hall. *International Law.* By W. E. Hall, M.A. *Fourth Edition.* 8vo. 22s. 6d.

—— *A Treatise on the Foreign Powers and Jurisdiction of the British Crown.* By W. E. Hall, M.A. 8vo. 10s. 6d.

Holland. *Elements of Jurisprudence.* By T. E. Holland, D.C.L. *Ninth Edition.* 8vo. 10s. 6d.

—— *The European Concert in the Eastern Question;* a Collection of Treaties and other Public Acts. Edited, with Introductions and Notes, by T. E. Holland, D.C.L. 8vo. 12s. 6d.

—— *Studies in International Law.* By T. E. Holland, D.C.L. 8vo. 10s. 6d.

—— *Gentilis, Alberici, De Iure Belli Libri Tres.* Edidit T. E. Holland, I.C.D. Small 4to, half-morocco, 21s.

—— *The Institutes of Justinian,* edited as a recension of the Institutes of Gaius, by T. E. Holland, D.C.L. *Second Edition.* Extra fcap. 8vo. 5s.

Holland and Shadwell. *Select Titles from the Digest of Justinian.* By T. E. Holland, D.C.L., and C. L. Shadwell, D.C.L. 8vo. 14s.

Also sold in Parts, in paper covers—
Part I. Introductory Titles. 2s. 6d.
Part II. Family Law. 1s.
Part III. Property Law. 2s. 6d.
Part IV. Law of Obligations (No. 1), 3s. 6d. (No. 2), 4s. 6d.

Ilbert. *The Government of India.* Being a Digest of the Statute Law relating thereto. With Historical Introduction and

Illustrative Documents. By Sir Courtenay Ilbert, K.C.S.I. 8vo, half-roan. 21s.

Jenks. *Modern Land Law.* By Edward Jenks, M.A. 8vo. 15s.

Markby. *Elements of Law considered with reference to Principles of General Jurisprudence.* By Sir William Markby, D.C.L. *Fifth Edition.* 8vo. 12s. 6d.

Moyle. *Imperatoris Iustiniani Institutionum Libri Quattuor,* with Introductions, Commentary, Excursus and Translation. By J. B. Moyle, D.C.L. *Third Edition.* 2 vols. 8vo. Vol. I. 16s. Vol. II. 6s.

—— *Contract of Sale in the Civil Law.* 8vo. 10s. 6d.

Pollock and Wright. *An Essay on Possession in the Common Law.* By Sir F. Pollock, Bart., M.A., and Sir R. S. Wright, B.C.L. 8vo. 8s. 6d.

Poste. *Gaii Institutionum Juris Civilis Commentarii Quattuor;* or, Elements of Roman Law by Gaius. With a Translation and Commentary by Edward Poste, M.A. *Third Edition.* 8vo. 18s.

Raleigh. *An Outline of the Law of Property.* By Thos. Raleigh, D.C.L. 8vo. 7s. 6d.

Sohm. *Institutes of Roman Law.* By Rudolph Sohm. Translated by J. C. Ledlie, B.C.L. With an Introductory Essay by Erwin Grueber, Dr. Jur., M.A. 8vo. 18s.

Stokes. *The Anglo-Indian Codes.* By Whitley Stokes, LL.D.
Vol. I. Substantive Law. 8vo. 30s.
Vol. II. Adjective Law. 8vo. 35s.
First and Second Supplements to the above, 1887–1891. 8vo. 6s. 6d.
Separately, No. 1, 2s. 6d.; No. 2, 4s. 6d.

3. HISTORY, BIOGRAPHY, ETC.

Adamnani *Vita S. Columbae.*
Ed. J. T. Fowler, D.C.L. Crown
8vo, half-bound, 8s. 6d. *net* (with
translation, 9s. 6d. *net*).

Aubrey. *'Brief Lives,' chiefly*
of Contemporaries, set down by John
Aubrey, between the Years 1669 *and*
1696. Edited from the Author's
MSS., by Andrew Clark, M.A., LL.D.
With Facsimiles. 2 vols. 8vo. 25s.

Baedae *Historia Ecclesiastica,*
etc. Edited by C. Plummer, M.A.
2 vols. Crown 8vo, 21s. *net.*

Bedford (W.K.R.). *The Blazon*
of Episcopacy. Being the Arms borne
by, or attributed to, the Arch-
bishops and Bishops of England
and Wales. With an Ordinary of
the Coats described and of other
Episcopal Arms. *Second Edition,*
Revised and Enlarged. With One
Thousand Illustrations. Sm. 4to,
buckram, 31s. 6d. *net.*

Boswell's *Life of Samuel*
Johnson, LL.D. Edited by G. Birkbeck Hill, D.C.L. In six volumes,
medium 8vo. With Portraits and
Facsimiles. Half-bound, 3l. 3s.

Bright. *Chapters of Early*
English Church History. By W.
Bright, D.D. *Third Edition. Revised*
and Enlarged. With a Map. 8vo. 12s.

Casaubon (Isaac). 1559–1614.
By Mark Pattison. 8vo. 16s.

Clarendon's *History of the*
Rebellion and Civil Wars in England.
Re-edited from a fresh collation of
the original MS. in the Bodleian
Library, with marginal dates and
occasional notes, by W. Dunn
Macray, M.A., F.S.A. 6 vols. Crown
8vo. 2l. 5s.

Hewins. *The Whitefoord*
Papers. Being the Correspondence
and other Manuscripts of Colonel
CHARLES WHITEFOORD and CALEB
WHITEFOORD, from 1739 to 1810.

Edited, with Introduction and
Notes, by W. A. S. Hewins, M.A.
8vo. 12. 6d.

Earle. *Handbook to the Land-*
Charters, and other Saxonic Documents.
By John Earle, M:A. Crown 8vo. 16s.

Earle and Plummer. *Two of*
the Saxon Chronicles, Parallel, with
Supp'ementary Extracts from the others.
A Revised Text, edited, with Intro-
duction, Notes, Appendices, and
Glossary, by Charles Plummer,
M.A., on the basis of an edition by
John Earle, M.A. 2 vols. Crown
8vo, half-roan.
Vol. I. Text, Appendices, and
    Glossary. 10s. 6d.
Vol. II. Introduction, Notes, and
    Index. 12s. 6d.

Freeman. *The History of*
Sicily from the Earliest Times.
Vols. I and II. 8vo, cloth, 2l. 2s.
Vol. III. The Athenian and
    Carthaginian Invasions. 24s.
Vol. IV. From the Tyranny of
    Dionysios to the Death of
    Agathoklês. Edited by Arthur
    J. Evans, M.A. 21s.

Freeman. *The Reign of*
William Rufus and the Accession of
Henry the First. By E. A. Freeman,
D.C.L. 2 vols. 8vo. 1l. 16s.

Gardiner. *The Constitutional*
Documents of the Puritan Revolution,
1628–1660. Selected and Edited
by Samuel Rawson Gardiner, D.C.L.
Second Edition. Crown 8vo. 10s. 6d.

Gross. *The Gild Merchant;*
a Contribution to British Municipal
History. By Charles Gross, Ph.D.
2 vols. 8vo. 24s.

Hastings. *Hastings and the*
Rohilla War. By Sir John Strachey,
G.C.S.I. 8vo, cloth, 10s. 6d.

Hill. *Sources for Greek*
History between the Persian and Pelopon-
nesian Wars. Collected and arranged
by G. F. Hill, M.A. 8vo. 10s. 6d.

Hodgkin. *Italy and her Invaders.* With Plates & Maps. 8 vols. 8vo. By T. Hodgkin, D.C.L.
Vols. I–II. *Second Edition.* 42s.
Vols. III–IV. *Second Edition.* 36s.
Vols. V–VI. 36s.
Vol. VII–VIII (*completing the work*). 24s.

Payne. *History of the New World called America.* By E. J. Payne, M.A. 8vo.
Vol. I, containing Book I, *The Discovery*; Book II, Part I, *Aboriginal America*, 18s.
Vol. II, containing Book II, *Aboriginal America* (concluded), 14s.

Johnson. *Letters of Samuel Johnson, LL.D.* Collected and Edited by G. Birkbeck Hill, D.C.L. 2 vols. half-roan, 28s.

—— *Johnsonian Miscellanies.* By the same Editor. 2 vols. Medium 8vo, half-roan, 28s.

Kitchin. *A History of France.* With Numerous Maps, Plans, and Tables. By G. W. Kitchin, D.D. In three Volumes. *New Edition.* Crown 8vo, each 10s. 6d.
Vol. I. to 1453. Vol. II. 1453–1624. Vol. III. 1624–1793.

Lewis (*Sir G. Cornewall*). *An Essay on the Government of Dependencies.* Edited by C. P. Lucas, B.A. 8vo, half-roan. 14s.

Lucas. *Introduction to a Historical Geography of the British Colonies.* By C. P. Lucas, B.A. With Eight Maps. Crown 8vo. 4s. 6d.

—— *Historical Geography of the British Colonies:*
Vol. I. The Mediterranean and Eastern Colonies (exclusive of India). With Eleven Maps. Crown 8vo. 5s.
Vol. II. The West Indian Colonies. With Twelve Maps. Crown 8vo. 7s. 6d.
Vol. III. West Africa. With Five Maps. Crown 8vo. 7s. 6d.

Vol. IV. South and East Africa. Historical and Geographical. With Ten Maps. Crown 8vo. 9s. 6d.
Also Vol. IV in two Parts—
Part I. Historical, 6s. 6d.
Part II. Geographical, 3s. 6d.

Ludlow. *The Memoirs of Edmund Ludlow, Lieutenant-General of the Horse in the Army of the Commonwealth of England,* 1625–1672. Edited by C. H. Firth, M.A. 2 vols. 8vo. 1l. 16s.

Machiavelli. *Il Principe.* Edited by L. Arthur Burd, M.A. With an Introduction by Lord Acton. 8vo. 14s.

Prothero. *Select Statutes and other Constitutional Documents. illustrative of the Reigns of Elizabeth and James I.* Edited by G. W. Prothero, M.A. Crown 8vo. *Second Edition.* 10s. 6d.

—— *Select Statutes and other Documents bearing on the Constitutional History of England, from* A.D. 1307 *to* 1558. By the same Editor. [*In Preparation.*]

Ramsay (Sir J. H.). *Lancaster and York.* A Century of English History (A.D. 1399–1485). 2 vols. 8vo. With Index, 37s. 6d.

Ramsay (W. M.). *The Cities and Bishoprics of Phrygia.* By W. M. Ramsay, D.C.L., LL.D.
Vol. I. Part I. The Lycos Valley and South-Western Phrygia. Royal 8vo. 18s. net.
Vol. I. Part II. West and West-Central Phrygia. 21s. net.

Ranke. *A History of England, principally in the Seventeenth Century.* By L. von Ranke. Translated under the superintendence of G. W. Kitchin, D.D., and C. W. Boase, M.A. 6 vols. 8vo. 63s. Revised Index, separately, 1s.

Rashdall. *The Universities of Europe in the Middle Ages.* By Hastings Rashdall, M.A. 2 vols. (in 3 Parts) 8vo. With Maps. 2l. 5s., net.

London: HENRY FROWDE, Amen Corner, E.C.

Smith's *Lectures on Justice,* *Police, Revenue and Arms.* Edited, with Introduction and Notes, by Edwin Cannan. 8vo. 10s. 6d. net.

—— *Wealth of Nations.* With Notes, by J. E. Thorold Rogers, M.A. 2 vols. 8vo. 21s.

Stephens. *The Principal Speeches of the Statesmen and Orators of the French Revolution,* 1789–1795. By H. Morse Stephens. 2 vols. Crown 8vo. 21s.

Stubbs. *Select Charters and other Illustrations of English Constitutional History, from the Earliest Times to the Reign of Edward I.* Arranged and edited by W. Stubbs, D.D., Lord Bishop of Oxford. *Eighth Edition.* Crown 8vo. 8s. 6d.

—— *The Constitutional History of England, in its Origin and Development. Library Edition.* 3 vols. Demy 8vo. 2l. 8s.

Also in 3 vols. crown 8vo, price 12s. each.

Stubbs. *Seventeen Lectures on the Study of Mediaeval and Modern History.* Crown 8vo. 8s. 6d.

—— *Registrum Sacrum Anglicanum.* An attempt to exhibit the course of Episcopal Succession in England. By W. Stubbs, D.D. Small 4to. *Second Edition.* 10s. 6d.

Swift (F. D.). *The Life and Times of James the First of Aragon.* By F. D. Swift, B.A. 8vo. 12s. 6d.

Vinogradoff. *Villainage in England.* Essays in English Mediaeval History. By Paul Vinogradoff, Professor in the University of Moscow. 8vo, half-bound. 16s.

Woodhouse. *Aetolia; its Geography, Topography, and Antiquities.* By William J. Woodhouse, M.A., F.R.G.S. With Maps and Illustrations. Royal 8vo, price 21s. net.

4. PHILOSOPHY, LOGIC, ETC.

Bacon. *Novum Organum.* Edited, with Introduction, Notes, &c., by T. Fowler, D.D. *Second Edition.* 8vo. 15s.

Berkeley. *The Works of* George Berkeley, D.D., formerly Bishop of Cloyne; including many of his writings hitherto unpublished. With Prefaces, Annotations, and an Account of his Life and Philosophy. By A. Campbell Fraser, Hon. D.C.L., LL.D. 4 vols. 8vo. 2l. 18s.

The Life, Letters, &c., separately, 16s.

Bosanquet. *Logic; or, the Morphology of Knowledge.* By B. Bosanquet, M.A. 8vo. 21s.

Butler. *The Works of Joseph Butler,* D.C.L., sometime Lord Bishop of Durham. Divided into sections,

with sectional headings, an index to each volume, and some occasional notes; also prefatory matter. Edited by the Right Hon. W. E. Gladstone. 2 vols. Medium 8vo. 14s. each.

Fowler. *The Elements of Deductive Logic, designed mainly for the use of Junior Students in the Universities.* By T. Fowler, D.D. *Tenth Edition,* with a Collection of Examples. Extra fcap. 8vo. 3s. 6d.

—— *The Elements of Inductive Logic, designed mainly for the use of Students in the Universities.* By the same Author. *Sixth Edition.* Extra fcap. 8vo. 6s.

—— *Logic;* Deductive and Inductive, combined in a single volume. Extra fcap. 8vo. 7s. 6d.

Fowler and **Wilson.** *The Principles of Morals.* By T. Fowler, D.D., and J. M. Wilson, B.D. 8vo, cloth, 14s.

Green. *Prolegomena to Ethics.* By T. H. Green, M.A. Edited by A. C. Bradley, M.A. *Fourth Edition.* Crown 8vo. 7s. 6d.

Hegel. *The Logic of Hegel.* Translated from the Encyclopaedia of the Philosophical Sciences. With Prolegomena to the Study of Hegel's Logic and Philosophy. By W. Wallace, M.A. *Second Edition, Revised and Augmented.* 2 vols. Crown 8vo. 10s. 6d. each.

Hegel's *Philosophy of Mind.* Translated from the Encyclopaedia of the Philosophical Sciences. With Five Introductory Essays. By William Wallace, M.A., LL.D. Crown 8vo. 10s. 6d.

Hume's *Treatise of Human Nature.* Edited, with Analytical Index, by L. A. Selby-Bigge, M.A. *Second Edition.* Crown 8vo. 8s.

—— *Enquiry concerning the Human Understanding, and an Enquiry concerning the Principles of Morals.* Edited by L. A. Selby-Bigge, M.A. Crown 8vo. 7s. 6d.

Leibniz. *The Monadology and other Philosophical Writings.* Translated, with Introduction and Notes, by Robert Latta, M.A., D.Phil. Crown 8vo. 8s. 6d.

Locke. *An Essay Concerning Human Understanding.* By John Locke. Collated and Annotated, with Prolegomena, Biographical, Critical, and Historic, by A. Campbell Fraser, Hon. D.C.L., LL.D. 2 vols. 8vo. 1l. 12s.

Lotze's *Logic,* in Three Books —of Thought, of Investigation, and of Knowledge. English Translation; edited by B. Bosanquet. M.A. *Second Edition.* 2 vols. Cr. 8vo. 12s.

—— *Metaphysic,* in Three Books—Ontology, Cosmology, and Psychology. English Translation; edited by B. Bosanquet, M.A. *Second Edition.* 2 vols. Cr. 8vo. 12s.

Martineau. *Types of Ethical Theory.* By James Martineau, D.D. *Third Edition.* 2 vols. Cr. 8vo. 15s.

—— *A Study of Religion :* its Sources and Contents. *Second Edition.* 2 vols. Cr. 8vo. 15s.

Selby-Bigge. *British Moralists.* Selections from Writers principally of the Eighteenth Century. Edited by L. A. Selby-Bigge, M.A. 2 vols. Crown 8vo. 18s.

Wallace. *Lectures and Essays* on Natural Theology and Ethics. By William Wallace, M.A., LL.D. Edited, with a Biographical Introduction by Edward Caird, M.A., Hon. D.C.L. 8vo, with a Portrait. 12s. 6d.

5. PHYSICAL SCIENCE, ETC.

Balfour. *The Natural History* of the Musical Bow. A Chapter in the Developmental History of Stringed Instruments of Music. Part I, Primitive Types. By Henry Balfour, M.A. Royal 8vo, paper covers. 4s. 6d.

Chambers. *A Handbook of Descriptive and Practical Astronomy.* By G. F. Chambers, F.R.A.S. *Fourth Edition,* in 3 vols. Demy 8vo.
Vol. I. The Sun, Planets, and Comets. 21s.
Vol. II. Instruments and Practical Astronomy. 21s.
Vol. III. The Starry Heavens. 14s.

De Bary. *Comparative Anatomy of the Vegetative Organs of the Phanerogams and Ferns.* By Dr. A. de Bary. Translated by F. O. Bower, M.A., and D. H. Scott, M.A. Royal 8vo. 1*l.* 2*s.* 6*d.*

—— *Comparative Morphology and Biology of Fungi, Mycetozoa and Bacteria.* By Dr. A. de Bary. Translated by H. E. F. Garnsey, M.A. Revised by Isaac Bayley Balfour, M.A., M.D., F.R.S. Royal 8vo, half-morocco, 1*l.* 2*s.* 6*d.*

—— *Lectures on Bacteria.* By Dr. A. de Bary. *Second Improved Edition.* Translated by H. E. F. Garnsey, M.A. Revised by Isaac Bayley Balfour, M.A., M.D., F.R.S. Crown 8vo. 6*s.*

Druce. *The Flora of Berkshire.* Being a Topographical and Historical Account of the Flowering Plants and Ferns found in the County, with short Biographical Notices. By G. C. Druce, Hon. M.A. Oxon. Crown 8vo, 16*s. net.*

Goebel. *Outlines of Classification and Special Morphology of Plants.* By Dr. K. Goebel. Translated by H. E. F. Garnsey, M.A. Revised by Isaac Bayley Balfour, M.A., M.D., F.R.S. Royal 8vo, half-morocco. 1*l.* 1*s.*

—— *Organography of Plants,* especially of the Archegoniatae and Spermaphyta. By Dr. K. Goebel. Authorized English Edition, by Isaac Bayley Balfour. M.A., M.D., F.R.S., Part I, General Organography. Royal 8vo, half-morocco, 12*s.* 6*d.*

Pfeffer. *The Physiology of Plants.* A Treatise upon the Metabolism and Sources of Energy in Plants. By Prof. Dr. W. Pfeffer. Second fully Revised Edition, translated and edited by Alfred J. Ewart, D.Sc., Ph.D., F.L.S. Part I. Royal 8vo, half-morocco, 28*s.*

Prestwich. *Geology—Chemical, Physical, and Stratigraphical.* By Sir Joseph Prestwich, M.A., F.R.S. In two Volumes. 3*l.* 1*s.*

Price. *A Treatise on the Measurement of Electrical Resistance.* By W. A. Price, M.A., A.M.I.C.E. 8vo. 14*s.*

Sachs. *A History of Botany.* Translated by H. E. F. Garnsey, M.A. Revised by I. Bayley Balfour, M.A., M.D., F.R.S. Crown 8vo. 10*s.*

Solms-Laubach. *Fossil Botany.* Being an Introduction to Palaeophytology from the Standpoint of the Botanist. By H. Graf zu Solms-Laubach. Translated by H. E. F. Garnsey, M.A. Revised by I. Bayley Balfour, M.A., M.D., F.R.S. Royal 8vo, half-morocco, 18*s.*

Warington. *Lectures on some of the Physical Properties of Soil.* By Robert Warington, M.A., F.R.S. With a Portrait of Prof. John Sibthorp. 8vo, 6*s.*

Biological Series.

I. *The Physiology of Nerve, of Muscle, and of the Electrical Organ.* Edited by Sir J. Burdon Sanderson, Bart., M.D., F.R.SS. L.&E. Medium 8vo. 1*l.* 1*s.*

II. *The Anatomy of the Frog.* By Dr. Alexander Ecker, Professor in the University of Freiburg. Translated, with numerous Annotations and Additions, by G. Haslam, M.D. Medium 8vo. 21*s.*

IV. *Essays upon Heredity and Kindred Biological Problems.* By Dr. A. Weismann. Authorized Translation. Crown 8vo.

Vol. I. Edited by E. B. Poulton, S. Schönland, and A. E. Shipley. *Second Edition.* 7*s.* 6*d.*

Vol. II. Edited by E. B. Poulton, and A. E. Shipley. 5*s.*